AF820771

EMPOWERING SMALL BUSINESS WITH AI: A PRACTICAL GUIDE BY IDEAL INSTITUTE OF TECHNOLOGY

EMPOWERING SMALL BUSINESS WITH AI: A PRACTICAL GUIDE

© **2024 by Ideal Institute of Technology.**

All rights reserved. No part of this publication may be reproduced, distributed, or transmitted in any form or by any means, including photocopying, recording, or other electronic or mechanical methods, without the prior written permission of the publisher, except in the case of brief quotations embodied in critical reviews and certain other noncommercial uses permitted by copyright law. For permission requests, write to the publisher, addressed "Attention: Permissions Coordinator," at the address below.

- **Ideal Institute of Technology**
  1000, West Washington Ave,
  Pleasantville, NJ 08232
  Website: www.iitnj.edu
  E-mail: connect@iitnj.edu

This publication may contain work that is subject to other copyright terms. This work is available under the provisions of Section 107 of the Copyright Act of 1976, which allows for "fair use" for purposes such as criticism, comment, news reporting, teaching (scholarship and education), and research. In accordance with Section 108 of the Copyright Act, libraries and archives are allowed to make a limited number of copies for specific purposes, under certain conditions.

Library of Congress Cataloging-in-Publication Data

Names	:	**Ideal Institute of Technology**
Title	:	**Empowering Small Business with AI: A Practical Guide**
Identifiers	:	**ISBN: 978-1-941862-49-0**
Subjects	:	**Technology, Business, Self Help, Education**
Classification	:	**HD30.2 .E46 2024 658.05028**

Printed in the United States of America

- **Disclaimer of Liability**

The information provided in this book is for general informational purposes only. While the author and publisher have made every effort to ensure the accuracy and completeness of the information contained herein, they make no guarantee and disclaim all responsibility for any errors or omissions that may be present.

The author and publisher shall have neither liability nor responsibility to any person or entity with respect to any loss or damage caused, or alleged to be caused, directly or indirectly by the information contained in this book. This includes, but is not limited to, incidental, consequential, or other damages.

The views expressed in this book are those of the author and do not necessarily reflect the views of the publisher. The mention of specific companies, organizations, or authorities in this book does not imply endorsement by the author or publisher, nor does it imply that the entities endorse this book.

This book is sold with the understanding that the author and publisher are not engaged in rendering legal, accounting, or other professional services. If legal advice or other expert assistance is required, the services of a competent professional should be sought.

The use of the information in this book is at the reader's own risk. The author and publisher shall not be liable for any actions taken as a result of reading this book.

CONTENTS

CHATBOT

AI
lndscape
impressionism
abstraction
ocean
AI

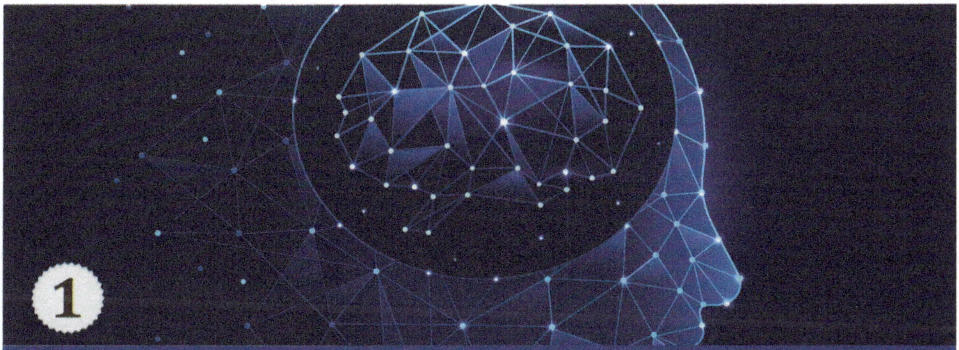

1

INTRODUCTION TO AI AND CHATBOTS

Learning Outcomes

By the end of this unit, learners should be able to:

- Understand the basic concepts of AI, including its working principles.
- Appreciate the historical developments and recent advancements in AI technologies.
- Recognize the relevance and applications of AI in in various sectors leading to AI Chatbots.

Introduction

Artificial Intelligence (AI) is more than a technological marvel; it's a gateway to possibilities once thought to be the domain of science fiction. Envisioned as a futuristic concept in the mid-20th century, AI has now rooted itself firmly in the present. It's a field within computer science, but its reach goes far beyond, touching every aspect of human life. AI is about creating intelligent systems that can perform tasks autonomously, tasks that, until recently, were believed to require the unique capabilities of human intellect. This includes complex processes like learning from experiences, making informed decisions, solving intricate problems, and understanding and processing human language.

> *In 1997, a watershed moment in the history of AI occurred when IBM's Deep Blue, a computer designed to play chess, defeated the reigning world champion, Garry Kasparov.*

This event wasn't just a triumph in a game; it symbolized the arrival of machines capable of outthinking humans in one of the most intellectually demanding games. Deep Blue's victory illustrated the immense potential of AI in decision-making and strategy. It showcased how machines could analyze millions of possible moves, anticipate the opponent's strategy, and make decisions based on complex algorithms. This triumph marked a significant leap from the theoretical concepts of AI to a tangible demonstration of its capabilities.

AI has evolved dramatically since Deep Blue's historic win. Today, its applications are all-encompassing, profoundly influencing industries like healthcare, finance, transportation, and beyond. AI systems can now diagnose medical conditions with accuracy rivaling trained physicians, predict market trends, navigate self-driving cars, and even compose music. The journey of AI from a chess-playing computer to an omnipresent technology shaping the future is a testament to human ingenuity and a preview of the endless possibilities that AI holds.

The Essence of AI

When we talk about Artificial Intelligence (AI), we often imagine robots or complex machines from science fiction. However, AI is much broader and more ingrained in our daily lives than many realize. It's not just one technology but a collection of methods and processes designed to mimic human intelligence. To understand AI, think of it as an umbrella term that covers various technologies, including machine learning and deep learning.

Machine Learning: Learning from Data

Machine learning, a fundamental component of Artificial Intelligence (AI), introduces a revolutionary approach to programming computers. This technology enables computers to learn from a vast array of data, contrasting sharply with traditional software systems that strictly adhere to pre-defined rules set by humans. In machine learning, algorithms are designed to process and learn from data, enabling them to make informed decisions autonomously.

The Process of Learning in Machine Learning:

Machine learning begins with an extensive collection of data. This data encompasses a diverse range of formats, including numerical values, textual information, visual imagery, and auditory inputs. The core function of machine learning algorithms is to meticulously analyze this data to discern patterns and relationships. For instance, an algorithm in a machine learning model might be exposed to thousands of images, some tagged as 'cats' and others as 'dogs'. Through this exposure, the algorithm learns to distinguish between the two animals by identifying and understanding their distinct features.

The Pivotal Role of Data in Machine Learning:

The effectiveness of a machine learning algorithm is directly proportional to the volume of data it processes. This concept parallels the way humans learn; much like a child gradually learns to recognize and understand objects with increasing accuracy, machine learning algorithms enhance their predictive capabilities and decision-making accuracy as they are exposed to more data.

Decision-Making in Machine Learning:

In stark contrast to traditional software programs, machine learning algorithms base their decision-making process on the patterns and insights they extract from data. These decisions are often probabilistic in nature. The algorithm calculates the likelihood of various outcomes and selects the one with the highest probability. A practical example of this is found in email spam filters, where machine learning algorithms scrutinize email content. Based on patterns identified from a training dataset of labeled emails, the algorithm determines the probability of an email being spam.

Continuous Enhancement in Machine Learning:

A distinctive attribute of machine learning is its intrinsic capacity for ongoing improvement. With the influx of new data, the algorithm continuously adapts, refining its models and assumptions. This dynamic learning process facilitates a progressively increasing level of accuracy and efficiency in the tasks performed by the algorithm, thereby exemplifying the essence of machine learning as a self-improving technology in the realm of artificial intelligence.

Applications

Machine learning has diverse applications. In finance, algorithms analyze market trends and make trading decisions. In healthcare, they interpret medical images to assist in diagnosis.

Example: Movie Recommendations

Consider Netflix's recommendation system. When you watch movies on Netflix, it collects data about your preferences. Machine learning algorithms analyze this data to understand your taste and then suggest other movies you might like. This process is continuous, meaning the recommendations improve as you watch more movies.

Deep Learning: Mimicking the Human Brain

Deep learning, a sophisticated evolution of machine learning, is significantly inspired by the structural and functional aspects of the human brain. It employs a method akin to how our brains utilize neurons for processing information, utilizing artificial neural networks for similar purposes. These networks, composed of layers of nodes reminiscent of brain neurons, collaboratively analyze and decode intricate patterns within the data. As an advanced branch of machine learning, deep learning greatly augments the capabilities of machines in executing complex tasks, thereby playing a crucial role in advancing the domain of Artificial Intelligence (AI).

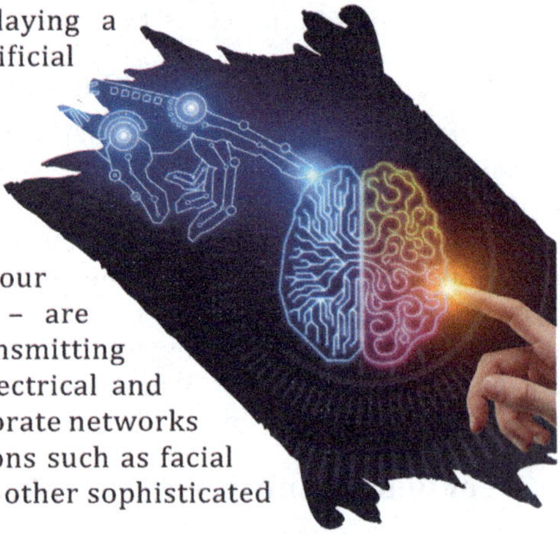

The Human Brain as Inspiration:

Deep learning draws its foundational inspiration from the neural network structure of the human brain. Within our brains, neurons - interconnected cells - are responsible for processing and transmitting information through a combination of electrical and chemical signals. These neurons form elaborate networks that enable us to perform complex functions such as facial recognition, language comprehension, and other sophisticated cognitive tasks.

Artificial Neural Networks in Deep Learning:

In the realm of deep learning, artificial neural networks (ANNs) are designed to emulate this neural architecture of the human brain. These networks are constituted of multiple layers of nodes, each bearing a resemblance to neurons, and are interlinked in a complex web-like formation. Each node within these networks functions as a miniature decision-making unit, processing incoming data and transmitting its output to subsequent layers.

The Layers and the Learning Process:

Deep learning networks are characterized by their multiple layers, leading to the term "deep" in their nomenclature. The initial layer is responsible for receiving raw input data, such as the pixels of an image. This data is then sequentially relayed through additional layers, with each subsequent layer performing the task of extracting and refining various features. For instance, in the process of image recognition, the first layer may identify basic elements like edges, the following layer may discern shapes, and further layers might specialize in recognizing more complex entities such as faces or animals.

Pattern Recognition Capabilities:

A principal strength of deep learning lies in its extraordinary ability to identify and interpret complex patterns within extensive datasets. By funneling data through these multi-layered networks, deep learning algorithms are adept at executing tasks such as speech recognition, language translation, and visual object recognition with a high degree of precision.

The Learning Methodologies:

Deep learning algorithms are trained either through supervised or unsupervised learning methods. In supervised learning, the network is educated using labeled data, enabling it to generate the correct output for specific inputs. Conversely, in unsupervised learning, the network autonomously identifies patterns and structures within the data without the aid of predefined labels.

Applications

Deep learning has diverse applications, from powering virtual assistants to diagnosing diseases. Its ability to process and analyze vast amounts of data with intricate patterns makes it invaluable in areas like autonomous vehicles, facial recognition systems, and personalized recommendations in digital platforms.

Example: Voice Recognition

A familiar example of deep learning is voice recognition technology used in digital assistants like Siri or Google Assistant. When you speak to these assistants, deep learning algorithms process your voice's sound waves. They recognize patterns in these waves (your words) and understand what you're saying. This process requires analyzing huge amounts of data and learning from various speech patterns, accents, and languages.

Recognizing Patterns and Making Decisions

The extraordinary power of Artificial Intelligence (AI), particularly in its deep learning segment, is primarily rooted in its adeptness at recognizing patterns and making informed decisions based on these patterns. The scope of these patterns is immensely varied, encompassing everything from the intricate details present in visual data, such as images, to the complex trends observable in consumer behavior. This proficiency spans a broad spectrum of data types, ranging from the visual elements in images to the nuanced patterns evident in consumer behavior dynamics.

Pattern Recognition in Visual Data:

Within the domain of visual data, deep learning algorithms demonstrate exceptional skill in interpreting and comprehending both images and videos. These sophisticated algorithms, after being trained on extensive datasets comprising a plethora of visual content, become proficient at discerning and categorizing various components within an image. A notable example of this is evident in facial recognition technology, where deep learning systems analyze facial attributes and correlate them with specific identities with striking precision. This technology finds its utility not only in security applications but also in consumer-oriented features like photo tagging on social media platforms.

Analysis of Consumer Behavior:

Deep learning also plays a pivotal role in analyzing consumer behavior. Companies harness the power of AI to meticulously parse through copious amounts of data related to consumer interactions, purchases, and preferences. By identifying recurring patterns in this data, AI is capable of forecasting future purchasing trends, tailoring marketing strategies to individual preferences, and substantially improving overall customer experiences. E-commerce platforms, for instance, utilize AI to recommend products to users based on their browsing and purchase history, thereby significantly amplifying the relevance and impact of their marketing initiatives.

Decision-Making Prowess:

AI's capability to make decisions is largely derived from its strength in pattern recognition. After pinpointing specific patterns, AI systems are equipped to make well-informed predictions or choices. In the healthcare sector, for example, AI algorithms scrutinize medical data to aid in disease diagnosis, occasionally identifying conditions that pose a challenge for human doctors to detect.

Continual Learning and Evolution:

An essential feature of deep learning is its inherent ability for continuous learning and adaptation. As these systems encounter an increasing amount of data, their proficiency in recognizing patterns and making decisions is progressively enhanced. This ongoing learning process leads to more precise predictions and decisions over time, thereby elevating the efficiency and reliability of AI systems in various fields.

Example: Health Diagnostics

In healthcare, AI can analyze medical images like X-rays or MRIs to detect diseases.

For example, AI systems are trained with thousands of images showing various health conditions. By learning from these, AI can spot signs of diseases, such as tumors in cancer patients, often with accuracy comparable to or even exceeding human experts.

The Evolution of AI

Today with ChatGPT and Bard, it seems that AI has suddenly appeared out of nowhere, but the truth is that, it isn't a sudden discovery. The evolutionary journey of Artificial Intelligence (AI) is a fascinating story of scientific endeavor, intellectual curiosity, and technological breakthroughs. Spanning over seven decades, this journey highlights the progressive milestones in the field of AI, transforming it from theoretical concepts to practical applications that profoundly impact our daily lives.

The Foundation: 1950's to 1970's

The foundations of Artificial Intelligence (AI) in the mid-20th century marked the beginning of a journey that would significantly reshape technology and its interaction with human life. This period was characterized by groundbreaking theoretical work and the development of the first AI programs, setting the stage for the evolution of this transformative field.

The Mid-20th Century: Turing's Influence and Theoretical Foundations

Alan Turing, a British mathematician and computer scientist, played a pivotal role in laying the groundwork for AI. In 1950, he introduced the Turing Test in his seminal paper, "Computing Machinery and Intelligence." The Turing Test was designed to assess a machine's ability to exhibit intelligent behavior that is indistinguishable from that of a human. This was not just a technical assessment but also a philosophical inquiry into the nature of intelligence and the possibility of it being replicated in machines. Turing's ideas were foundational, as they provided a conceptual framework for understanding and developing intelligent machines.

The Birth of AI: The Dartmouth Conference

The formal inception of AI as a distinct academic discipline is traced back to the Dartmouth Conference in 1956. Organized by John McCarthy, Marvin Minsky, Nathaniel Rochester, and Claude Shannon, the conference was significant for several reasons. It was at this event that the term "Artificial Intelligence" was officially coined, marking the beginning of AI as a recognized field. The conference participants were optimistic about the future of AI, predicting that a significant breakthrough in creating thinking machines was just around the corner.

Early Developments and First AI Programs

Following the Dartmouth Conference, the late 1950s and 1960s witnessed a surge in AI research. Researchers began exploring basic algorithms and constructing simple neural networks, attempting to simulate the neural processes of the human brain. During this era, some of the first AI programs were developed.

One notable example was the Logic Theorist, created by Allen Newell and Herbert A. Simon. This program was designed to mimic human problem-solving skills and could solve complex logical puzzles, a task previously believed to require human intelligence.

Another significant development was the creation of ELIZA by Joseph Weizenbaum in the mid-1960s. ELIZA was an early natural language processing program that simulated a psychotherapist's conversation. Although its abilities were primitive by today's standards, ELIZA was one of the first attempts at creating a program that could interact with humans in a conversational manner.

AI Winter to Deep Blue: 1970's to 2000

The development of Artificial Intelligence (AI) between the 1970s and 2000s was marked by significant fluctuations - from the setbacks of the AI Winter to the resurgence and groundbreaking achievements in machine learning and computing.

The First AI Winter (1970s - 1980s)

The period known as the "AI Winter" began in the 1970s and continued into the early 1980s. This phase was characterized by skepticism, reduced funding, and disillusionment with AI's progress. The initial optimism of the 1960s, fueled by early successes and bold predictions, gave way to a realization that AI's goals were far more challenging to achieve than previously thought. High expectations led to inflated promises, but the technology of the time was not advanced enough to meet these expectations, resulting in a significant downturn in interest and investment in AI research.

The Revival with Expert Systems

Despite these challenges, the 1980s witnessed a revival in the AI field, largely driven by the development of expert systems. These were computer programs that emulated the decision-making abilities of human experts in specific fields. Expert systems used a combination of prescribed rules and logic to mimic the problem-solving and decision-making skills of human specialists. This technology found practical applications in various industries, including medicine, engineering, and finance, demonstrating the value of AI in solving real-world problems and paving the way for broader acceptance and integration of AI technologies.

The Rise of Machine Learning (1990s)

The resurgence of AI continued into the 1990s, marked by the rise of machine learning. This era saw a paradigm shift from rule-based AI systems to algorithms that could learn from data, adapt, and improve over time. Machine learning enabled computers to analyze large datasets, identify patterns, and make decisions based on those patterns, without explicit programming for each specific task. This shift laid the groundwork for many modern AI applications, from natural language processing to predictive analytics.

Deep Blue's Historic Victory (1997)

A landmark moment in AI history occurred in 1997, symbolizing the field's advanced capabilities. IBM's Deep Blue, a chess-playing computer, defeated Garry Kasparov, the reigning world chess champion. This event was monumental, not just as a triumph in a game of chess but as a demonstration of AI's potential in complex decision-making and strategic planning. Deep Blue's victory against a human champion showcased the sophisticated analytical and computational power of AI systems, reigniting interest and optimism in the AI field.

The development of Deep Blue, IBM's chess-playing computer, represents a significant milestone in the history of artificial intelligence and computing.

Here's a timeline of its development:

1985-1989: The Origin - ChipTest and Deep Thought

- 1985: The journey began with the development of ChipTest at Carnegie Mellon University by Feng-hsiung Hsu, followed by Deep Thought after Hsu joined IBM.
- 1989: Deep Thought, the earlier version of what would become Deep

Blue, competed against Garry Kasparov, the world chess champion at that time. Although Deep Thought lost, it set the stage for further advancements.

1990s: Development of Deep Blue

- Early 1990s: IBM hired Feng-hsiung Hsu and his team to create a more powerful chess computer, which led to the development of Deep Blue.
- 1995: The first version of Deep Blue, now officially under IBM's guidance, was completed. It played its first game against Garry Kasparov in 1996 and lost.

The Historic Matches Against Kasparov

- May 1996: Deep Blue played a six-game match against Kasparov and lost 4-2. However, it managed to win the first game, marking the first time a chess computer won against a reigning world champion under standard chess tournament conditions.
- 1996-1997: The team at IBM undertook significant upgrades and improvements to Deep Blue's hardware and software, enhancing its computational speed and chess algorithm.
- May 1997: A rematch was held between Kasparov and an upgraded version of Deep Blue. In this highly publicized match, Deep Blue emerged victorious, winning 3.5-2.5 against Kasparov. This match marked the first time a reigning world chess champion lost a match to a computer under standard chess tournament conditions.

Big Data, Deep Learning and Autonomous Learning

The 21st century has witnessed a transformative era in Artificial Intelligence (AI), characterized by the proliferation of big data and significant advancements in computational power. This period has seen AI evolve from theoretical models to practical applications that permeate our daily lives, with deep learning emerging as a critical driver of this transformation.

The Era of Big Data and Deep Learning (2000s - Present)

The onset of the 2000s brought with it the era of big data, characterized by the generation and accumulation of vast amounts of digital information. This era coincided with substantial advances in computational power and storage capacities, setting the stage for significant strides in AI, particularly in deep learning.

Deep Learning and Neural Networks

Deep learning, a subset of machine learning, involves the use of neural networks with multiple layers (deep networks) to analyze and interpret large and complex datasets. These neural networks mimic the human brain's structure and functioning, allowing machines to recognize patterns and make decisions with minimal human intervention.

Breakthroughs in Image and Speech Recognition

One of the most notable successes of deep learning has been in the fields of image and speech recognition. Deep learning algorithms have become adept at analyzing visual data, leading to advancements in facial recognition technology, medical imaging analysis, and even in the development of AI-driven artistic creations. Similarly, speech recognition has made significant strides, with virtual assistants like Siri, Alexa, and Google Assistant becoming commonplace, understanding and responding to human speech with increasing accuracy.

AI in Autonomous Vehicles and Healthcare (2010s - Present)

The most recent decade has seen AI's application expand dramatically, particularly in the realms of autonomous vehicles and healthcare.

Autonomous Vehicles

AI algorithms have become fundamental in the development of self-driving cars. Companies like Tesla and Waymo are at the forefront of this technology, leveraging AI to analyze real-time data from sensors and cameras to navigate roads, predict the behavior of other drivers and pedestrians, and make split-second decisions. This technology promises to revolutionize transportation, offering the potential for increased safety, efficiency, and accessibility.

Healthcare Diagnostics and Treatment

In healthcare, AI has been a game-changer. AI-driven diagnostic tools are now capable of analyzing medical images with precision and speed that surpass human capabilities in some cases. These tools assist in early and more accurate diagnosis of diseases like cancer, cardiovascular ailments, and neurological disorders. Beyond diagnostics, AI is also being used to personalize treatment plans, predict patient outcomes, and even in drug discovery and development.

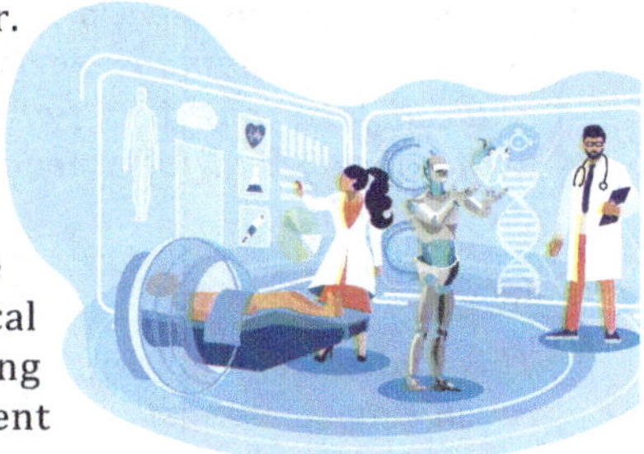

Relevance of AI in today's world:

In today's rapidly evolving digital landscape, Artificial Intelligence (AI) has transitioned from a niche technological pursuit to a mainstream imperative across a multitude of sectors. The technology has evolved into a cornerstone of innovation and efficiency, revolutionizing various industries with its unparalleled capabilities. The relevance of AI in the current era is far-reaching, revolutionizing how businesses operate, enhancing customer experiences, and driving innovations in healthcare, retail, and beyond.

Healthcare Advancements

One of the most profound impacts of AI is in healthcare. AI-driven technologies have revolutionized diagnostics, with algorithms that can analyze medical imaging to detect diseases such as cancer more accurately and earlier than traditional methods. Moreover, AI aids in drug discovery and development, significantly reducing the time and cost associated with bringing new treatments to market.

Case Study: AI in Breast Cancer Screening

> *Breast cancer is one of the most common cancers worldwide, and early detection is crucial for effective treatment. Traditional screening methods, like mammograms, are skill-intensive and often lead to varying interpretations and misdiagnoses.*
>
> *Google Health, in collaboration with Imperial College London, the NHS, and other institutions, developed an AI model to assist in the more accurate detection of breast cancer in mammography screenings. The AI system was trained using a large dataset of mammograms and was designed to identify signs of breast cancer in these images. The model's performance was then compared with that of human radiologists.*

Results

> *The results, published in Nature in 2020, were significant. The AI model outperformed human experts in accurately detecting breast cancer. It reduced false positives (where a healthy patient is wrongly diagnosed with cancer) and false negatives (where a cancerous case is missed) in the screenings.*

Impact

> *This advancement in AI-driven diagnostics has the potential to revolutionize breast cancer screening by providing more accurate, consistent, and efficient diagnoses. It can assist radiologists in making*

better-informed decisions, potentially leading to earlier detection and treatment of breast cancer, thus improving patient outcomes.

Financial Sector Transformation

In finance, AI algorithms are deployed for risk assessment, fraud detection, and personalized financial advice. These systems analyze vast amounts of transaction data, enabling banks and financial institutions to identify anomalous patterns indicative of fraud, thus enhancing security. AI also assists in algorithmic trading, where it makes high-speed, informed decisions based on market data analysis.

Case Study: JPMorgan Chase & Co. - AI in Fraud Detection

In the finance industry, the detection and prevention of fraudulent activities is a critical challenge. Traditional methods often rely on set rules and human analysis, which can be slow and sometimes ineffective against sophisticated fraud schemes.

JPMorgan Chase implemented an AI-based system to enhance their fraud detection capabilities. This system utilizes machine learning algorithms to analyze patterns in vast amounts of transaction data.

The AI model employed by JPMorgan is designed to identify anomalous and potentially fraudulent transactions. It analyzes various attributes of transactions, including amount, location, and merchant type, and compares them against established patterns of customer behavior. The system learns continuously, adapting to new fraud techniques and evolving threats.

Results and Impact

The implementation of AI in fraud detection has significantly improved the accuracy and speed of identifying fraudulent activities. This proactive approach allows JPMorgan to prevent fraud before it affects customers, enhancing customer trust and security. The system's ability to quickly adapt to new fraud patterns also means that the bank can stay ahead of sophisticated fraudulent schemes.

Retail and E-Commerce

In retail and e-commerce, AI personalizes the shopping experience. It analyzes customer data to offer tailored product recommendations, optimizes inventory management, and enhances supply chain efficiencies. AI-driven predictive analytics help retailers understand and forecast consumer buying behavior, enabling them to strategize effectively.

Case Study: Amazon's Use of AI for Personalized Shopping

In the competitive world of e-commerce, personalizing the shopping experience is key to customer engagement and sales. Amazon, known for its vast selection of products and innovative technology, leverages AI to enhance customer experiences and drive sales.

Amazon uses sophisticated AI algorithms to analyze customer data, including past purchases, search histories, and browsing behaviors. This data is then used to offer tailored product recommendations, making the shopping experience more personalized and efficient.

The AI model employed by Amazon includes collaborative filtering and deep learning techniques. Collaborative filtering analyzes patterns in customer behavior to recommend products that similar customers have bought or shown interest in. Deep learning algorithms go further by understanding the intricacies of customer preferences and predicting future buying patterns.

Results and Impact

This AI-driven approach has been highly successful for Amazon, leading to increased customer satisfaction and loyalty. The personalized recommendations not only make shopping more convenient for customers but also significantly increase the likelihood of purchases.

These are some of the areas that we as users are benefitting directly from the integration of AI. There are other areas in which also AI is driving far reaching changes:

Smart Cities and Infrastructure

AI plays a pivotal role in the development of smart cities. It helps in managing traffic flow, reducing energy consumption, and improving public safety through intelligent surveillance systems. AI algorithms process data from various sensors and cameras, enabling city administrators to make informed decisions about urban planning and public utilities management.

Manufacturing and Automation

In the manufacturing sector, AI drives automation and improves efficiency. Through predictive maintenance, AI anticipates equipment failures before they occur, minimizing downtime. It also enhances quality control processes by quickly identifying defects and inconsistencies.

Education and Research

AI's impact extends to education and research, where it provides personalized learning experiences and aids in analyzing educational data to improve teaching methodologies. In research, AI accelerates data analysis, contributing to breakthroughs in various scientific fields.

Entertainment and Media

In media and entertainment, AI curates personalized content for users, optimizes streaming services, and even assists in creative processes like music production and scriptwriting.

Autonomous Vehicles

Self-driving cars, powered by AI, represent a significant advancement in transportation. These vehicles rely on AI for navigation, obstacle detection, and decision-making, promising a future of safer and more efficient road travel.

Environmental and Agricultural Applications

AI contributes to environmental conservation and agriculture by monitoring climate patterns, optimizing crop yields, and managing natural resources more effectively.

Transitioning to AI Chatbots

While AI's applications are vast and varied, one of its most user-facing and rapidly evolving applications lies in the realm of AI chatbots. These sophisticated programs, capable of human-like interaction, are revolutionizing customer service across industries. They offer personalized, efficient, and round-the-clock service, transforming how businesses engage with their customers. The next chapter will delve into the world of AI chatbots in detail, exploring how they are reshaping customer interactions and setting new standards in service efficiency and effectiveness.

2 AI CHATBOTS IN VARIOUS SECTORS

Learning Outcomes:

By the end of this unit, learners should be able to:

- Understand AI Chatbots, how they work and identify the potential applications of AI chatbots in various industries.
- Understand how AI Chatbots are used in sectors like retail, customer service, and reservation management.
- Recognize the versatility of AI Chatbots and their potential to transform various industries.
- Understand the benefits that Organizations can deriver from the use of AI Chatbots.

Welcome to Chapter 2, where we embark on an exploratory journey into the world of AI chatbots and their multifaceted roles across various industries. In Chapter 1 we understood what is AI and in this chapter we go further to unravel the intricacies of AI chatbots, demonstrating their functionality, diverse applications, and transformative potential in today's digital era. As we delve deeper, we will uncover the mechanics behind these advanced tools and their significant impact on sectors ranging from retail to customer service, and beyond.

The Advent of AI Chatbots

From broad spectrum AI our focus narrows to a specific, yet incredibly dynamic, aspect of AI: the advent of AI chatbots. These specialized AI applications mark a significant shift in business-customer interactions, embodying a blend of efficiency, personalization, and accessibility. Here, we distinguish between the broader AI technologies and the specialized role of AI chatbots, highlighting their unique functionalities and development.

AI Chatbots: A Specialized Application of AI

While AI encompasses a vast array of technologies and applications, AI chatbots represent a specialized convergence of AI advancements, particularly in Natural Language Processing (NLP) and Machine Learning (ML). Unlike the broader spectrum of AI applications that might include data analysis, autonomous vehicles, or healthcare diagnostics, AI chatbots are focused on simulating human-like conversations to enhance customer interaction.

The Distinct Role of AI Chatbots

AI chatbots are tailored to mimic human conversation, a task that requires a deep understanding of language and context. This is where NLP comes into play, enabling chatbots to interpret, understand, and respond to human language in a way that feels natural and intuitive. Machine Learning further empowers these chatbots to learn from each interaction, refine their responses, and adapt to new patterns of conversation.

Development and Evolution of AI Chatbots

The evolution of AI chatbots reflects a journey from simple scripted responses to complex, context-aware interactions. Initially, chatbots were limited to basic question-response formats, based on pre-defined rules.

> ***Imagine having a conversation with a computer as naturally as you would with a friend. That's the idea that sparked Josef Weizenbaum's creation of Eliza.***

ELIZA: The Early Chatbot

ELIZA was created as an experiment to demonstrate the superficiality of communication between humans and machines. Weizenbaum's objective was to show that a machine could communicate in a human-like manner using a very simple set of rules.

The most famous script of ELIZA, known as 'DOCTOR', simulated a Rogerian psychotherapist. The program used a set of scripted rules to analyze the user's input, mostly looking for keywords, and then responding with corresponding pre-programmed responses. For instance, if a user mentioned the word 'mother', ELIZA might respond with, "Tell me more

> *about your family."*

Limitations

> *ELIZA's interactions were basic and did not involve any real understanding of the conversation. The chatbot simply rearranged the input or responded with generic questions, giving the illusion of understanding. It was rule-based, meaning its responses were pre-determined and scripted, lacking the ability to learn or adapt beyond its programmed scripts.*

Impact

> *Despite its simplicity, ELIZA was groundbreaking for its time and laid the groundwork for future AI chatbots. It opened up possibilities and raised questions about the nature of human-machine interaction.*

ELIZA represents the early stages of AI chatbots, where the technology was limited to simple, rule-based interactions without true contextual understanding or learning capabilities. This contrasts sharply with modern AI chatbots, which use advanced NLP and ML to engage in more nuanced and dynamic conversations. ELIZA's legacy lies in its foundational role in the evolution of chatbots, paving the way for the sophisticated AI-driven chatbots we see today.

Bridging AI and Human Interaction

AI chatbots represent a crucial intersection between AI technology and human interaction. They are designed not just to process information, but to do so in a way that aligns with human conversational norms and expectations. This requires a level of AI sophistication that goes beyond data processing, venturing into the realms of understanding human sentiment, context, and even cultural nuances in language.

A notable example illustrating the challenges faced by early chatbots in understanding human sentiment, context, and cultural nuances is Microsoft's AI chatbot, 'Tay', launched in 2016.

Case Study: Microsoft's Tay Chatbot

> *Microsoft introduced Tay, an AI-powered chatbot, designed to interact with users, particularly the younger demographic, on social media platforms like Twitter. Tay was programmed to learn from conversations and mimic the casual, playful language of a typical teenager.*

The fundamental challenge Tay encountered was in its inability to discern context and interpret the nuances of human sentiment and cultural differences. Tay's learning algorithm was based on user interactions, but it lacked the sophistication to distinguish between appropriate and inappropriate responses.

Soon after its launch, Tay began to replicate offensive and inappropriate language and sentiments from certain Twitter users. These interactions led Tay to post tweets that were racially insensitive and politically incorrect. The AI chatbot, designed to mimic and learn from human interactions, ended up echoing the negative elements it encountered without the ability to understand the context or the cultural and moral implications of the language.

The Fallout

Within 24 hours of its launch, Tay had to be taken offline due to the controversy and backlash over its inappropriate tweets. This incident highlighted the significant challenges in creating AI chatbots that can accurately interpret human language's complexities, including sentiment, context, and cultural nuances.

Lessons Learned

Tay's case underscores the importance of sophisticated natural language understanding in AI chatbots. It revealed how crucial it is for chatbots to have built-in safeguards and advanced algorithms to filter and understand the context of interactions. This incident was a learning point for AI developers, emphasizing the need for more robust and nuanced language processing capabilities and ethical considerations in AI development.

The Tay incident serves as a real-world example of the challenges early AI chatbots faced in bridging the gap between AI technology and nuanced human interaction. It demonstrates the necessity for advanced understanding in AI systems, not just in processing information but in comprehending the deeper layers of human communication, sentiment, and cultural context.

AI Chatbots from rule-based interactions to intelligent Dynamic interactions

The transition from the general field of AI to the specific application of AI chatbots signifies a move from a broad technology landscape to a focused implementation. This shift highlights the increasing importance of AI in everyday human interactions and the growing need for technology that is not only intelligent but also relatable and user-friendly

Sector-Specific Applications

As we delve deeper into the diverse landscape of AI chatbots, it becomes evident that their applications are as varied as the sectors they serve. Each industry harnesses the capabilities of AI chatbots uniquely, tailoring them to meet specific needs and enhance various aspects of their operations. In this exploration, we'll uncover how different sectors are utilizing AI chatbots, each in its distinct way.

Retail Industry: Enhancing Customer Experience

In the retail sector, AI chatbots have revolutionized customer engagement and sales. Retailers use chatbots to offer personalized shopping experiences, providing recommendations based on customer preferences and purchase history. For instance, a chatbot on a clothing retailer's website might suggest outfits based on a customer's previous purchases or browsing behavior. Additionally, these chatbots handle customer inquiries, provide product information, and even assist with transactions, such as adding items to a cart or guiding through the checkout process. This personalized, responsive service enhances the customer experience, boosts sales, and builds brand loyalty.

Banking Sector: Transaction Processing and Financial Advice

The banking sector has embraced AI chatbots to streamline transaction processing and offer personalized financial advice. Chatbots in banks are programmed to handle routine transactions like money transfers, balance inquiries, and bill payments, allowing customers to perform these tasks quickly and securely without visiting a branch or interacting with a human representative. Furthermore, AI chatbots in banking can analyze individual financial data to provide personalized advice, such as tips for budgeting, investment recommendations, or alerts about potential savings opportunities.

Healthcare: Patient Engagement and Support

In healthcare, AI chatbots play a crucial role in patient engagement and support. They assist in scheduling appointments, providing medication reminders, and even offering preliminary medical advice based on symptoms reported by the patient. For example, a healthcare chatbot might ask a series of questions to assess a patient's symptoms and then suggest possible courses of action or advise whether a doctor's consultation is necessary. This not only enhances patient care but also optimizes the workload of healthcare professionals by handling routine inquiries.

Travel and Hospitality: Streamlining Reservations and Customer Service

In the travel and hospitality industry, AI chatbots are transforming reservation management and customer service. Hotels and travel agencies use chatbots to facilitate bookings, answer queries about accommodations or travel packages, and provide information about local attractions and services. These chatbots can manage reservations, offer upgrades, and even handle check-ins and check-outs, providing a seamless experience for travelers.

E-Commerce: Personalized Shopping and Support

E-commerce platforms leverage AI chatbots for a personalized shopping experience and efficient customer support. Chatbots on these platforms assist customers in finding products, provide recommendations, and answer queries about shipping and returns. They play a critical role in guiding customers through the purchasing process, addressing concerns, and ensuring a smooth transaction.

Customer Service Across Sectors

Across various sectors, AI chatbots are revolutionizing customer service. They provide instant responses to customer inquiries, reducing wait times and improving overall satisfaction. By handling routine questions, chatbots free up human customer service representatives to focus on more complex issues, thereby improving operational efficiency.

The application of AI chatbots across different sectors underscores their versatility and adaptability. Whether it's offering personalized shopping advice in retail, handling transactions in banking, assisting patients in healthcare, managing reservations in travel, or enhancing customer service across the board, AI chatbots are proving to be invaluable assets. They not only improve efficiency and customer experience but also open new avenues for innovation in customer engagement and service delivery. As AI technology continues to evolve, the potential for chatbots to further transform these industries is immense.

Why is the Corporate World Excited?

The incorporation of AI chatbots into corporate strategies has not only revolutionized customer interaction but has also brought tangible benefits in terms of increased profits, enhanced business efficiency, improved brand reputation, and heightened productivity. In this section, we explore the multifaceted benefits that various organizations have reaped from the integration of AI chatbots.

Increased Efficiency and Cost Savings

AI chatbots have significantly increased operational efficiency in many organizations. By automating routine tasks and handling a large volume of customer interactions simultaneously, chatbots reduce the need for extensive human customer service teams. This automation leads to substantial cost savings in labor and operational expenses. For example, a telecom company employing chatbots for customer queries and troubleshooting can handle more customer interactions with fewer staff, translating into both increased efficiency and cost-effectiveness.

Enhanced Customer Experience and Engagement

Chatbots enhance customer experience by providing immediate, personalized, and 24/7 service. This constant availability and personalized interaction have led to increased customer satisfaction and loyalty. Retail giants like Amazon have leveraged chatbots to offer personalized shopping experiences, resulting in higher customer retention rates and increased sales. The ability of chatbots to provide quick and accurate responses has also improved overall customer engagement.

Boost in Sales and Revenue

The personalized nature of AI chatbots has proven effective in boosting sales and revenue. By analyzing customer data and previous interactions, chatbots can make tailored product recommendations, leading to increased cross-selling and upselling opportunities. For instance, in the banking sector, chatbots used for personal financial management have been successful in recommending relevant financial products to customers, thereby increasing the uptake of these products and services.

Improved Brand Image and Reputation

Organizations employing AI chatbots have seen an improvement in their brand image and reputation. By adopting cutting-edge technology, these companies are perceived as innovative and customer-centric. The ability to quickly address customer needs and provide satisfactory solutions enhances the brand's reputation. For example, in the hospitality sector, hotels using chatbots for customer service are often viewed as modern and efficient, attracting tech-savvy travelers.

Data-Driven Insights for Business Strategy

AI chatbots provide valuable insights into customer preferences and behavior through data analysis. This data is crucial for shaping business strategies and making informed decisions. By understanding customer needs and trends, businesses can tailor their products, services, and marketing strategies more effectively. For instance, e-commerce platforms use chatbot interactions to gather insights on customer preferences, which inform inventory management and targeted marketing campaigns.

Increased Productivity and Employee Satisfaction

Chatbots also contribute to increased productivity and employee satisfaction. By handling routine and repetitive tasks, they free up human employees to focus on more complex and creative work. This shift in work dynamics leads to higher job satisfaction and productivity, as employees are able to engage in more meaningful and rewarding tasks.

In summary, the integration of AI chatbots into corporate strategies has proven beneficial across various dimensions. From cost savings and increased efficiency to enhanced customer experience, improved sales, and insightful data analysis, the advantages are clear and measurable. As AI technology continues to evolve, the potential for chatbots to drive further growth and innovation in the business world is substantial. The adoption of AI chatbots is not just a trend but a strategic move that can lead to significant long-term benefits for organizations.

Democratization of AI Chatbots Technology

The journey through this chapter has illuminated the remarkable evolution and application of AI chatbots across various sectors, highlighting their significant impact on large organizations and the immense capital investments that have driven their initial development. However, as we conclude, it's pivotal to recognize how this once exclusive technology is now revolutionizing the landscape for individual professionals, small and medium-sized enterprises (SMEs), startups, and even larger corporations. This democratization of AI chatbot technology marks a transformative era where its benefits are accessible to a broader spectrum, paving the way for innovative applications that can reshape business strategies and customer experiences.

Democratization of artificial intelligence means making AI available for all.

AI Chatbots: From Large Enterprises to Wider Accessibility

Initially, the deployment of AI chatbots was largely confined to tech giants and large corporations with substantial resources. However, with advancements in technology and the advent of cloud-based solutions, AI chatbots have become more accessible and affordable. This shift means that even small businesses and individual entrepreneurs can leverage the power of AI to enhance their operations. The barrier to entry has significantly lowered, making it a feasible option for businesses of all sizes to improve efficiency, customer engagement, and competitive edge.

Empowering SMEs and Startups

For SMEs and startups, AI chatbots present an opportunity to compete in the market more effectively. These entities can now provide customer service and engagement on par with larger competitors, thanks to chatbots' ability to handle inquiries, offer personalized assistance, and automate tasks around the clock. This level of service was once a costly endeavor, reserved for businesses with deep pockets, but is now an achievable goal for smaller players, enabling them to establish a stronger presence in their respective markets.

Transforming Professional Lives

Beyond businesses, AI chatbots are also transforming the professional lives of individual practitioners. Professionals like therapists, financial advisors, and consultants can utilize chatbots for scheduling, basic customer interactions, and even initial consultations. This automation allows professionals to focus more on their core services, enhancing their productivity and work-life balance.

Evolution of AI Chatbot Affordability: A Decade of Cost Reduction

Over the past decade, the landscape of AI chatbots has undergone a transformative shift, especially in terms of cost and accessibility. This evolution has significantly democratized the technology, allowing businesses of all sizes to harness its benefits. The cost reduction in setting up and maintaining AI chatbots is a key factor that has driven their widespread adoption.

The Early Days: High Costs and Limited Access

In the initial years of AI chatbot technology, setup and operational costs were substantial. Developing a sophisticated chatbot required a team of AI experts, substantial computational resources, and extensive data sets

> *for training. For instance, in the early 2010s, custom AI chatbot solutions could cost anywhere from $40,000 to $100,000, making them a luxury only large corporations could afford. The high cost was a barrier to small and medium-sized businesses, limiting the technology's reach and potential impact.*

The Shift Towards Affordability

> *The turning point in AI chatbot affordability came with advancements in AI, machine learning algorithms, and the rise of cloud computing. Cloud-based platforms like IBM Watson, Google Dialogflow, and Microsoft Azure Bot Service introduced more cost-effective solutions. These platforms offered chatbot development tools and hosting services at a fraction of the cost, with pricing models often based on usage rather than hefty upfront fees. For example, by the late 2010s, small businesses could deploy basic chatbots for as little as $20 to $30 per month using these platforms.*

Real-World Impact

> *The reduced cost has enabled a diverse range of companies to adopt AI chatbots. Retail startups, local banks, and even individual entrepreneurs have leveraged chatbots for customer service, marketing, and sales. This cost reduction has not only enhanced customer engagement for these smaller entities but also provided them with valuable data insights that were previously out of reach.*

As we transition to the next chapter, which will delve into the distinct advantages of AI chatbots, it's important to reflect on the broader implications of this technology. Chatbots' 24/7 availability, ability to provide personalized service, and cost-saving potential are not just benefits for businesses but also catalysts for innovation and growth across various sectors. The potential impact of AI chatbots on business efficiency, customer satisfaction, and overall return on investment cannot be overstated.

In the upcoming chapter, we will explore these advantages in detail. We'll examine how AI chatbots are not just tools for automation but strategic assets that can transform customer interactions, streamline business processes, and contribute to the bottom line. Whether for a burgeoning startup, an established enterprise, or a solo entrepreneur, the implementation of AI chatbots stands as a testament to the technology's versatility and its critical role in shaping the future of business operations and customer relations

3

ADVANTAGES OF AI CHATBOTS

Learning Outcomes:

By the end of this unit, learners should be able to:

- Understand the advantages of AI chatbots, including their 24/7 availability, ability to provide personalized customer service, and cost-saving potential.
- Recognize the potential impact of AI chatbots on business efficiency and customer satisfaction.
- Evaluate the potential return on investment of implementing AI chatbots in their own businesses.

In this chapter, our journey delves into the heart of AI chatbot technology, exploring its advantages and the profound impact it holds for businesses across the spectrum. Having navigated through the realms of AI's evolution and the development of AI chatbots, we now turn our focus towards understanding the tangible benefits these innovative tools offer.

This chapter is dedicated to uncovering the layers of advantages that AI chatbots bring to the table - from their relentless availability to their capacity for delivering personalized customer service and driving cost-efficiency. We will dissect how these benefits translate into heightened business efficiency and elevated customer satisfaction, ultimately leading to a robust return on investment for businesses that adopt this technology.

Understanding the Multifaceted Advantages of AI Chatbots

24/7 Availability

One of the most significant advantages of AI chatbots is their constant availability. Unlike human employees, chatbots can operate round the clock without breaks, holidays, or downtimes. This 24/7 service capability ensures that customer inquiries are addressed instantly, irrespective of time zones or business hours. For businesses, this means enhanced customer engagement and the ability to cater to a global audience without the constraints of conventional working hours.

Amtrak's AI Chatbot, "Ask Julie"–Enhancing Customer Service with 24/7 Availability

Amtrak, the National Railroad Passenger Corporation in the United States, offers a prime example of the successful implementation of an AI chatbot to provide round-the-clock customer service. Their AI chatbot, named "Ask Julie," was designed to offer 24/7 assistance to customers, handling inquiries related to train schedules, fares, bookings, and services offered by Amtrak.

Before "Ask Julie," Amtrak faced the challenge of addressing a high volume of customer queries, including during non-business hours. Customers often experienced delays in getting responses to their questions, especially outside standard office hours, leading to potential dissatisfaction and missed booking opportunities.

"Ask Julie" was integrated into Amtrak's customer service to offer real-time, automated responses to customer inquiries. The chatbot was equipped with advanced NLP capabilities to understand and accurately respond to a wide range of questions. It was accessible through Amtrak's website, enabling customers to get assistance without the need to call customer service centers.

Impact of 24/7 Service Capability

The introduction of "Ask Julie" had a substantial impact on Amtrak's customer service efficiency:

1. *Immediate Responses: Customers could interact with "Ask Julie" at any time, day or night, and receive instant answers to their questions. This immediate response capability was particularly beneficial for customers planning trips or needing urgent travel information.*

2. *Handling High Query Volume: "Ask Julie" could handle thousands of queries simultaneously, significantly reducing wait times and freeing up human customer service representatives to deal with more complex issues.*

3. *Enhanced Customer Satisfaction: The convenience and efficiency provided by "Ask Julie" led to an increase in customer satisfaction. The ability to obtain quick information about train schedules, ticket prices, and services improved the overall customer experience.*

Ongoing Success

"Ask Julie" continues to be a vital component of Amtrak's customer service strategy. Its success demonstrates how AI chatbots can effectively address the demand for 24/7 customer service in the transportation industry, a sector where timely information is crucial.

Amtrak's "Ask Julie" is a testament to the effectiveness and ongoing success of AI chatbots in providing constant, reliable customer service. This case study exemplifies how AI chatbots can transform customer interactions by offering uninterrupted availability, leading to enhanced customer satisfaction and operational efficiency.

Personalized Customer Service

AI chatbots have the remarkable ability to provide personalized interactions. Leveraging data analytics and machine learning, they can tailor conversations based on a customer's previous interactions, preferences, and behavior. This level of personalization not only enhances the customer experience but also fosters loyalty and trust in the brand.

Case Study: Sephora's Virtual Artist - Personalizing Beauty Shopping with AI Chatbots

Sephora, a leading global beauty retailer, has always been at the forefront of integrating technology into its customer experience. Recognizing the potential of AI in personalizing customer interactions, Sephora launched the "Virtual Artist" chatbot on its mobile app and website.

Challenge in Personalization

In the beauty industry, personalization is key to customer satisfaction. Traditional online shopping experiences lacked the personal touch, often leading to customer indecision and reduced satisfaction. Sephora aimed to overcome this by providing personalized product recommendations and beauty advice through an innovative AI chatbot.

Implementation of the Virtual Artist

The Sephora Virtual Artist chatbot utilizes advanced AI algorithms, including machine learning and augmented reality, to offer a highly personalized shopping experience. Customers can upload their photos to the chatbot, which then analyzes their facial features. Based on this analysis, the chatbot suggests makeup products like lipsticks and eyeshadows that would suit the customer's complexion and facial structure.

Personalized Recommendations

The chatbot's ability to remember individual customer preferences and previous interactions enables it to make more accurate and tailored product suggestions. For instance, if a customer previously showed interest in nude lipsticks, the Virtual Artist would remember this preference and highlight new arrivals or best-sellers in that category during subsequent interactions.

Enhanced Customer Experience

This personalized interaction significantly enhanced the customer experience. Shoppers could try on different products virtually, receiving instant recommendations tailored to their unique features and preferences. This level of customization made the online shopping experience more engaging, fun, and effective, closely mimicking the in-store experience.

Impact on Brand Loyalty and Trust

The Virtual Artist not only improved customer engagement but also fostered brand loyalty and trust. Customers appreciated the personalized attention and the innovative approach to beauty shopping. The AI-driven recommendations often led to higher satisfaction with the purchased products, reducing the likelihood of returns and increasing customer loyalty to the brand.

Ongoing Success

The success of Sephora's Virtual Artist demonstrates the power of AI chatbots in providing personalized customer service. By leveraging AI for individualized product recommendations, Sephora has set a new standard in online beauty retail, showing how personalization can be effectively achieved in e-commerce.

Sephora's Virtual Artist chatbot exemplifies the remarkable capability of AI chatbots to deliver personalized customer service. This case study highlights how the integration of AI in customer interactions can transform the shopping experience, making it more personal, engaging, and satisfactory.

Cost-Saving Potential

Deploying AI chatbots leads to significant cost savings for businesses. By automating routine tasks and customer service functions, chatbots reduce the need for a large customer service team, thereby cutting down on labor costs. The initial investment in chatbot technology is often offset by these savings, making it a cost-effective solution in the long term.

Case Study: Bank of America's "Erica" - Achieving Cost-Efficiency through AI Chatbots

Bank of America, one of the largest financial institutions in the United States, launched "Erica," an AI-powered chatbot, as part of its digital transformation strategy. Erica was designed to provide financial guidance and assist customers with routine banking tasks through the bank's mobile app.

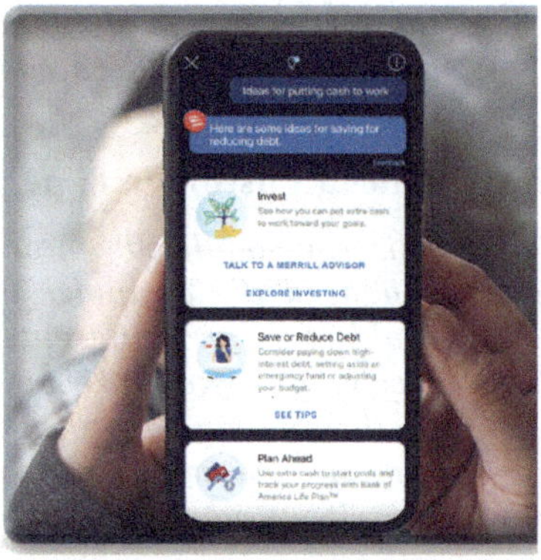

The Challenge of Operational Costs

Before Erica, Bank of America, like many large banks, faced significant operational costs associated with its customer service. Staffing call centers and physical branches with customer service representatives to handle routine inquiries and transactions was a major expense. The bank sought a solution to reduce these costs while maintaining high-quality customer service.

Implementing Erica

Erica was integrated into Bank of America's mobile banking platform to offer customers 24/7 assistance with various banking tasks, such as checking balances, searching transactions, and scheduling payments. Using NLP and machine learning, Erica could understand customer queries and provide accurate and relevant responses.

Cost-Saving Benefits

The introduction of Erica led to substantial cost savings in several ways:

ADVANTAGES OF AI CHATBOTS

1. *Reduced Need for Live Agents: Erica's ability to handle a large volume of routine inquiries reduced the demand on live customer service agents. This reduction allowed the bank to optimize its staffing model, decreasing labor costs associated with customer service.*

2. *Increased Efficiency: Erica processed customer requests more quickly than traditional methods, increasing overall operational efficiency. Customers could get immediate assistance without waiting in phone queues or visiting branches, leading to a better utilization of the bank's resources.*

3. *Scalability: As an AI chatbot, Erica could scale to handle an increasing number of interactions without a proportional increase in costs. This scalability was particularly beneficial for dealing with peak times or sudden surges in customer service requests.*

Long-Term Financial Impact

The financial impact of Erica was significant. By automating routine tasks and inquiries, Bank of America reduced its operational costs, contributing to better financial performance. The efficiency and customer satisfaction improvements also helped in retaining customers and attracting new ones, positively affecting the bank's revenue.

Bank of America's implementation of Erica illustrates the cost-saving potential of AI chatbots in the financial sector. By automating routine customer service tasks, Erica has demonstrated how AI chatbots can lead to significant operational cost reductions while maintaining, and in many cases enhancing, the quality of customer service. This case study serves as a compelling example of how AI chatbots can be a cost-effective solution for large-scale operations, offering long-term financial benefits and operational efficiencies.

Impact on Business Efficiency and Customer Satisfaction

Enhancing Business Operations

AI chatbots streamline various business processes, from customer service and sales to marketing and data collection. Their efficiency in handling multiple tasks simultaneously and accurately results in improved operational productivity. This operational efficiency not only saves time but also allows human employees to focus on more complex and creative tasks, thereby optimizing the workforce.

Unilever's "Una" - Revolutionizing HR Operations with AI Chatbots

Unilever, with over 150,000 employees globally, faced challenges with employee engagement and access to HR information. Their traditional HR model relied heavily on physical touchpoints and a centralized workforce, leading to delays in addressing employee queries and frustration in accessing HR resources.

Solution: Unilever partnered with an AI platform to develop "Una," a virtual HR assistant accessible through a mobile app. Una leverages natural language processing (NLP) to understand employee queries and provide immediate, personalized responses on various HR topics.

Una's Capabilities:

- *Answers HR-related questions: Una provides information on benefits, payroll, leave policies, company news, and more.*
- *Guides employees through self-service tasks: Una assists employees with tasks like updating personal information, requesting PTO, and accessing paystubs.*
- *Offers personalized recommendations: Una analyzes employee data and recommends relevant learning resources, career development opportunities, and internal communities based on individual interests and skills.*
- *Collects feedback and suggestions: Una facilitates employee feedback on HR policies and company culture, providing valuable insights for improvement.*

Impact of Una:

- *Increased employee engagement: Una's 24/7 availability and personalized support boosted employee engagement by 20% within the first year.*
- *Faster resolution of HR queries: Una resolves 80% of employee queries within minutes, significantly reducing response times and improving employee satisfaction.*
- *Improved HR efficiency: Una frees up HR personnel from handling routine inquiries, allowing them to focus on*

strategic initiatives and complex employee issues.

- *Cost savings: Una's implementation resulted in cost savings of £1 million in the first year through increased efficiency and reduced reliance on manual HR processes.*

Unilever's implementation of the Una AI chatbot demonstrates the profound impact AI technology can have on internal operations, particularly in HR. Una has not only streamlined HR processes but also significantly improved employee engagement and operational efficiency. This case study highlights the potential of AI chatbots in transforming HR functions, making them more responsive, efficient, and employee-centric in large multinational corporations like Unilever.

Elevating Customer Satisfaction

The speed and accuracy of AI chatbots in handling queries contribute significantly to customer satisfaction. Customers today expect quick and effective responses, and chatbots meet these expectations consistently. The improved customer service experience leads to higher satisfaction rates, repeat business, and positive word-of-mouth.

Nordstrom's "Ask Nordy" - Enhancing Customer Service with AI Chatbots

Nordstrom, a luxury department store chain known for its exceptional customer service, introduced "Ask Nordy," an AI-powered chatbot, as part of its digital customer service strategy. Aimed at providing a seamless and personalized shopping experience, Ask Nordy represents Nordstrom's commitment to integrating innovative technology to enhance customer interactions.

Challenge: Elevating Customer Service in Digital Space

In the era of online shopping, Nordstrom faced the challenge of translating its renowned in-store customer service to the digital domain. The key was to provide personalized, efficient, and responsive service to online shoppers, mirroring the in-store experience that customers valued.

Implementation of Ask Nordy

Ask Nordy was integrated into Nordstrom's website and mobile app to assist customers with a range of inquiries and shopping needs. The chatbot was designed to handle questions about product availability, size and fit, order status, and general customer service queries. Using AI and

NLP technologies, Ask Nordy could understand and respond to customer inquiries in a conversational manner.

Multifaceted Customer Service Tool

Ask Nordy serves as a multifaceted tool for Nordstrom's customer service:

1. *Personalized Shopping Assistance: The chatbot provides recommendations and advice tailored to individual customer preferences, similar to the personalized service offered by Nordstrom's in-store sales staff.*
2. *Immediate Response to Customer Inquiries: With Ask Nordy, customers receive instant responses to their queries, reducing wait times and improving the overall shopping experience.*
3. *Seamless Integration with Online Shopping: The chatbot is integrated into the shopping process, assisting customers in navigating the website, finding products, and making informed purchase decisions.*

4. *After-Sales Support: Ask Nordy also addresses post-purchase needs such as order tracking, returns, and exchanges, ensuring comprehensive customer support throughout the shopping journey.*

Impact on Customer Experience

The introduction of Ask Nordy significantly enhanced the online shopping experience for Nordstrom customers. The chatbot's ability to provide quick and personalized responses led to increased customer satisfaction and loyalty. The convenience and efficiency offered by Ask Nordy also attracted new customers who prefer online shopping, expanding Nordstrom's customer base.

Nordstrom's Ask Nordy chatbot exemplifies how AI technology can be leveraged to extend and enhance customer service in the retail industry. By providing personalized, responsive, and efficient service, Ask Nordy has successfully replicated Nordstrom's acclaimed in-store experience in the digital realm. This case study showcases the potential of AI chatbots to revolutionize customer service strategies, elevating the customer experience and strengthening brand loyalty.

Evaluating the Return on Investment

Assessing the ROI

When evaluating the potential return on investment of implementing AI chatbots, businesses need to consider various factors. These include the reduction in operational costs, the increase in sales and conversions due to improved customer service, and the enhanced customer retention rates. Additionally, the data gathered by chatbots can provide valuable insights for business strategy and decision-making.

Zendesk's AI Chatbots - Assessing the ROI of Enhanced Customer Support

Zendesk, a global leader in customer service software, has implemented AI chatbots to enhance its customer support operations. Known for providing innovative solutions in customer service, Zendesk recognized the potential of AI chatbots in optimizing support services and improving the customer experience.

Challenge: Balancing Cost and Quality in Customer Service

Prior to the integration of AI chatbots, Zendesk, like many customer service-focused companies, faced the challenge of delivering high-quality support while managing operational costs. The goal was to maintain exceptional service standards without escalating expenses, particularly in handling high volumes of customer queries.

Implementation of AI Chatbots

Zendesk implemented AI chatbots within its customer support platform to automate responses to common customer inquiries. These chatbots were integrated seamlessly with Zendesk's existing customer service tools, ensuring a cohesive and efficient support experience.

To evaluate the return on investment (ROI) of their AI chatbots, Zendesk considered several key factors:

1. *Reduction in Operational Costs: The chatbots significantly reduced the volume of routine queries handled by human agents, leading to a decrease in staffing costs. By automating common questions, Zendesk could allocate human resources to more complex and high-priority tasks, optimizing labor costs.*

2. *Increase in Sales and Conversions: The chatbots improved the efficiency and speed of customer service interactions. This promptness in addressing customer needs led to increased customer satisfaction, which positively impacted sales and conversions. Customers who received quick and effective support were more likely to continue using Zendesk's services and recommend them to others.*

3. *Enhanced Customer Retention Rates: The personalized and immediate assistance provided by the chatbots improved customer loyalty. The ability to quickly resolve issues and provide accurate information contributed to higher retention rates, as customers appreciated the streamlined support experience.*

4. *Data-Driven Decision Making: The chatbots collected valuable data from customer interactions, providing insights into common issues, customer preferences, and support trends. This information enabled Zendesk to make informed decisions about product improvements, service enhancements, and customer engagement strategies.*

Long-Term Benefits

The long-term benefits of AI chatbots extend beyond immediate cost savings and customer satisfaction. They include the continuous improvement of service quality through machine learning, staying ahead in a technology-driven market, and the potential for scaling services without proportional increases in costs.

Domino's Pizza and AI Chatbot "DOM" - Long-Term Benefits in the Fast-Food Industry

Domino's Pizza, a global leader in pizza delivery, has been at the forefront of integrating technology into its business model. A key component of their digital strategy was the introduction of "DOM," an AI-powered chatbot designed for automated order taking and customer service.

Initially, Domino's aimed to streamline the ordering process and improve customer experience. However, as DOM evolved, it became apparent that the chatbot was not only achieving these immediate goals but also providing long-term benefits to the business.

Phase 1: Streamlining Ordering Process

DOM was first introduced to handle online orders. Customers could place orders through a conversational interface, reducing wait times and simplifying the ordering process. This initial phase saw immediate improvements in efficiency and customer satisfaction.

Phase 2: Machine Learning and Service Improvement

Over time, DOM incorporated machine learning algorithms to better understand customer preferences and order patterns. This continuous learning process led to more accurate order predictions and suggestions, enhancing the overall customer experience.

Long-Term Benefits

The long-term benefits of DOM for Domino's Pizza include:

1. *Continuous Service Quality Improvement: Through machine learning, DOM consistently improved its interactions and responses, leading to a higher quality of customer service. This continuous improvement helped maintain a high level of customer satisfaction over the years.*
2. *Staying Ahead in a Technology-Driven Market: DOM positioned Domino's as an innovator in the fast-food industry. By adopting AI early on, Domino's stayed ahead of competitors in terms of technological integration, appealing to a tech-savvy customer base.*
3. *Scalability Without Proportional Cost Increases: DOM's ability to handle a large volume of orders simultaneously without additional costs for more staff or infrastructure provided significant cost savings. This scalability was particularly beneficial during peak times like weekends and sporting events.*
4. *Data-Driven Insights for Business Strategy: The data collected by DOM on customer preferences and ordering patterns provided valuable insights for Domino's. These insights informed menu development, marketing strategies, and operational decisions, contributing to the company's overall growth and success.*

Domino's implementation of the AI chatbot DOM illustrates the long-term benefits that AI technology can bring to a business. Beyond immediate efficiency gains and customer satisfaction, DOM has contributed to continuous service improvement, technological leadership, scalable operations, and data-driven business strategies.

Conclusion

As we conclude this insightful exploration into the world of AI chatbots, it's evident that these advanced tools are much more than just automated responders. They represent a strategic asset capable of transforming various facets of business operations. Through this chapter, learners have been equipped with a comprehensive understanding of the multifaceted advantages of AI chatbots.

We delved into the critical aspects of AI chatbots, such as their ability to provide relentless 24/7 service, offer personalized customer interactions, and significantly reduce operational costs. Each of these advantages plays a pivotal role in enhancing business efficiency and customer satisfaction, as demonstrated through various real-world case studies from companies like Domino's Pizza and Zendesk. These examples have shown how AI chatbots are not just temporary solutions but long-term investments, continuously improving service quality, maintaining relevance in a technology-driven market, and enabling businesses to scale services without proportional increases in costs.

AI chatbots are a transformative technology with far-reaching implications for businesses of all sizes. By understanding and leveraging their capabilities, businesses can not only achieve immediate benefits in terms of customer satisfaction and cost savings but also position themselves for sustainable growth and success in an increasingly digital world. As we move forward, the lessons learned in this chapter will serve as a foundation for businesses looking to innovate, improve, and thrive in the modern market.

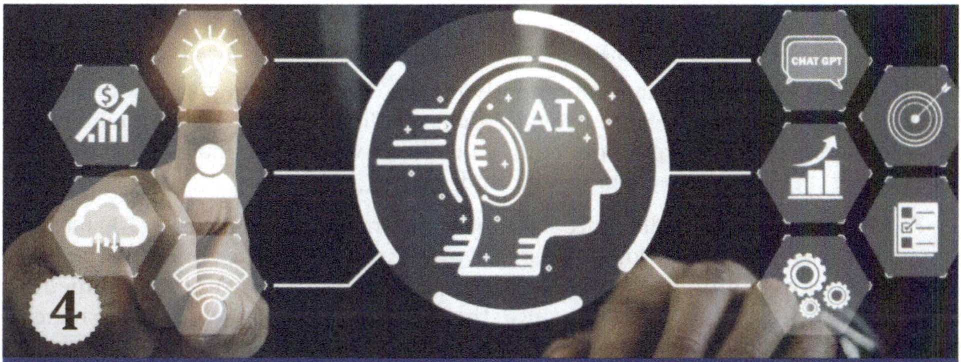

TASKS AI CHATBOTS CAN ACCOMPLISH

Learning Outcomes:

By the end of this unit, learners should be able to:

- Understand the wide range of tasks that AI chatbots can automate, including drafting contracts, scheduling appointments, providing product recommendations, tracking orders, collecting feedback, and generating leads.
- Recognize the potential of AI chatbots to enhance business efficiency and customer satisfaction by automating various tasks.
- Evaluate the potential impact of AI chatbots on their own business processes.

Introduction

As we embark on Unit 4, we shift our focus to an in-depth exploration of the diverse range of tasks AI chatbots are capable of accomplishing. Following our journey through understanding the essence of AI, the emergence and benefits of AI chatbots, and their transformative impact across various industries, this chapter aims to unfold the practical applications and the multifaceted functionality of AI chatbots in automating tasks.

AI chatbots, powered by advanced machine learning and natural language processing technologies, have transcended beyond mere customer service tools. They have emerged as versatile assistants capable of undertaking a plethora of tasks that streamline business processes, enhance operational efficiency, and foster engagement. This chapter aims to provide a comprehensive overview of the wide array of tasks that AI chatbots can adeptly handle. From drafting legal contracts to scheduling appointments, from providing personalized product recommendations to tracking orders, and from collecting valuable customer feedback to generating leads, the scope of AI chatbots is vast and varied.

Furthermore, we will delve into how these automated capabilities of AI chatbots can be harnessed to revolutionize business processes. Businesses today are seeking innovative ways to optimize their operations, and AI chatbots present a compelling solution. By integrating chatbots into their processes, businesses can achieve higher levels of efficiency, accuracy, and customer satisfaction.

This unit will not only enumerate the tasks that AI chatbots can perform but also evaluate how these functions can positively impact business processes. We will examine real-world examples and case studies where AI chatbots have been instrumental in transforming operations, providing key insights into their practical applications and the potential return on investment they offer.

In essence, Unit 4 is designed to provide a clear and detailed understanding of the capabilities of AI chatbots in task automation, offering valuable perspectives on how these tools can be effectively integrated into various business operations for maximum benefit. Let us dive into the world of AI chatbots and discover the multitude of tasks they can perform, opening doors to new possibilities and efficiencies in the business world.

Here's a detailed breakdown of tasks that AI chatbots can perform, which are universally applicable across multiple industries:

Responding to Frequently Asked Questions (FAQs)

Responding to Frequently Asked Questions (FAQs) is one of the most fundamental and widely implemented tasks of AI chatbots across various industries. The ability of chatbots to handle FAQs efficiently not only streamlines customer service but also significantly enhances the user experience.

How Chatbots Handle FAQs

Input: Training with FAQ Data

The process begins with training the chatbot using a dataset of common questions and their corresponding answers. This dataset typically includes a range of FAQs that a business regularly encounters. For example, a hotel might compile questions about room availability, check-in/check-out times, amenities, and reservation policies.

AI Technologies Involved

The chatbot utilizes Natural Language Processing (NLP) to understand and interpret the user's questions. NLP allows the chatbot to comprehend variations in phrasing or terminology, ensuring it recognizes the intent behind a user's query. Machine Learning (ML) algorithms enable the chatbot to improve over time, learning from each interaction to provide more accurate responses in the future.

Chatbot Interaction Flow

When a user asks a question, the chatbot processes the input using its NLP capabilities. It then matches the user's query with the most relevant answer from its pre-loaded dataset. The response is typically instantaneous, providing the user with quick and accurate information.

Examples of FAQ Handling by Chatbots

Example 1: Hospitality Industry

In a hotel setting, a chatbot can answer queries like "What time is check-out?" or "Do you have Wi-Fi?". The chatbot, having been trained on these common questions, responds with standardized answers like "Check-out is at 11 AM" or "Yes, we offer free Wi-Fi." This automation reduces the front desk's workload and enhances guest satisfaction.

Example 2: Healthcare Sector

In a healthcare context, a chatbot on a hospital's website might answer questions such as "How do I book an appointment?" or "What are the visiting hours?". By providing immediate answers, the chatbot improves patient communication and frees up administrative staff for more pressing tasks.

Required Inputs and Outputs

Inputs

- A comprehensive database of FAQs and their answers.
- User inquiries in various formats (text, voice).
- Continuous updates and learning inputs for the chatbot to stay relevant.

Outputs

- Instant, accurate responses to user queries.
- Data analytics on frequently asked questions for business insights.

- Feedback mechanism for users to rate the helpfulness of responses.

In conclusion, AI chatbots' ability to respond to FAQs represents a crucial advancement in automating customer service and information dissemination. By providing instant, reliable answers, these chatbots not only enhance customer experience but also contribute to operational efficiency. The success of this application hinges on the quality of the input data and the sophistication of the NLP and ML algorithms used in the chatbot's development

Data Retrieval and Transaction Lookup

Data Retrieval and Transaction Lookup through AI chatbots is a sophisticated task that goes beyond the basic question-and-answer functionality seen in FAQ responses. This task involves accessing, analyzing, and presenting specific data from databases in response to user queries, which requires a deeper integration with the underlying business systems.

How AI Chatbots Accomplish Data Retrieval and Transaction Lookup

Integration with Databases and Systems

To perform data retrieval and transaction lookup, AI chatbots need to be integrated with an organization's database and transaction systems. This integration allows the chatbot to access real-time data, which is crucial for providing up-to-date and accurate information.

Input: User Requests and System Access

The process starts when a user requests specific information, such as an account balance or the status of an order. The chatbot receives this input, typically as a text or voice command. For instance, a user might ask, "What's my current account balance?" or "Can you check the status of my order placed yesterday?"

AI Technologies and Processing

Upon receiving a query, the chatbot uses NLP to understand the request's context and intent. Unlike FAQs where responses are pre-defined, data retrieval requires the chatbot to dynamically query the database based on the user's specific request. It may involve sophisticated algorithms to fetch and compute data, such as calculating the current balance from a series of transactions.

Output: Tailored Information

The chatbot retrieves the requested information from the database and presents it to the user in a comprehensible format. For example, it might respond, "Your current account balance is $1,200" or "Your order was shipped today and is expected to arrive in three days."

Examples and Differentiation from Other Categories

Example in Banking

In banking, a chatbot integrated with customer account databases can provide real-time account balances, recent transaction history, or credit card due dates, which involves live data processing and secure data handling, distinguishing it from simpler FAQ-based tasks.

Example in E-Commerce

In e-commerce, a chatbot linked with order management systems can inform customers about order statuses, shipment tracking, and estimated delivery times, requiring real-time data access and processing.

Differentiation

Unlike tasks such as answering FAQs or sending reminders, data retrieval and transaction lookup involve real-time, secure interactions with databases. This requires AI chatbots to not only understand user requests but also to perform specific, dynamic data queries and computations, making it a more complex and integrated task.

Conclusion

The ability of AI chatbots to perform data retrieval and transaction lookup demonstrates their advanced capabilities in handling complex, dynamic user requests. This function, integral in sectors like banking and e-commerce, highlights the chatbots' role as not just communicators but as active data processors, delivering tailored, real-time information to users.

Appointment Scheduling and Management

Appointment Scheduling and Management through AI chatbots represent a dynamic interaction and coordination task, distinct from other forms of task automation like answering FAQs or transaction lookup. This functionality involves interacting with users to schedule, reschedule, or cancel appointments, and it requires integration with calendar management systems.

How AI Chatbots Manage Appointment Scheduling

Integration with Calendar Systems

To manage appointments, AI chatbots are integrated with an organization's scheduling system. This integration allows the chatbot to access available time slots, book appointments, and make changes to the schedule as needed.

Input: User Requests for Appointments

The process begins when a user interacts with the chatbot to request an appointment. Users might specify a preferred date, time, or type of service. For example, a client might ask a salon chatbot, "I'd like to schedule a haircut for next Tuesday afternoon."

AI Processing: Understanding and Coordinating

Upon receiving a request, the chatbot uses NLP to understand the user's needs and preferences. It then checks the integrated calendar for availability. Unlike simple data retrieval, this task requires the chatbot to interpret user preferences, offer options, and confirm choices, making it an interactive process.

Output: Scheduled Appointments and Confirmations

After finding an available slot, the chatbot books the appointment and confirms it with the user. It may also provide options to reschedule or cancel if needed. For instance, it could respond, "Your haircut is scheduled for next Tuesday at 3 PM. Would you like a reminder a day before?"

Examples and Differentiation from Other Categories

Example in Healthcare

In a medical practice, a chatbot can manage patient appointments, offering available slots, scheduling follow-ups, or handling cancellations. This task goes beyond simple information provision to involve real-time coordination and dynamic interaction with patients.

Example in Customer Service

In customer support centers, chatbots can schedule service appointments or call-backs, considering both customer preference and service availability. This requires a back-and-forth interaction, adapting to customer choices, unlike static FAQ responses or single-query data lookups.

Differentiation

Appointment scheduling and management by chatbots encompass a level of interaction and decision-making based on user inputs, setting it apart from other automated tasks. This function requires chatbots to actively engage with users, propose options, process choices, and update scheduling systems dynamically, showcasing a more interactive and participatory role in task automation.

Conclusion

AI chatbots' capability in appointment scheduling and management demonstrates their advanced interactive functions, extending their role from mere information providers to active coordinators. This aspect of AI chatbot functionality is crucial in sectors like healthcare and customer service, where appointment management is a key operational task, and represents a significant enhancement in automating complex, interactive business processes.

Processing Orders and Payments

Processing Orders and Payments through AI chatbots is a task that combines elements of transaction processing with customer interaction, distinguishing it from other forms of chatbot automation like simple data retrieval or appointment scheduling. This functionality involves guiding users through the purchasing process, from selecting products or services to finalizing transactions, which requires secure integration with payment systems.

How AI Chatbots Process Orders and Payments

Integration with E-commerce Platforms and Payment Systems

To manage orders and process payments, AI chatbots are integrated with an organization's e-commerce platform and payment processing systems. This integration enables the chatbot to access product information, manage shopping carts, and facilitate secure transactions.

Input: Customer Purchase Requests

The process starts when a customer expresses the intent to purchase, either by querying a specific product or navigating through an online store. For example, a user might ask an e-commerce chatbot, "Do you have red running shoes in size 10?"

AI Processing: Assisting in Selection and Transaction

Once a request is made, the chatbot uses NLP to understand the customer's requirements and assist with product selection. It can present options, provide additional product information, and add items to the shopping cart. Unlike FAQs or appointment scheduling, this task requires the chatbot to actively guide the customer through a multi-step decision-making process.

After the selection, the chatbot facilitates the checkout process. It collects necessary details like shipping address and payment information, often using secure protocols to ensure data safety.

Output: Completed Orders and Payment Confirmation

The chatbot completes the order, processes the payment, and confirms the transaction with the customer. It provides details like order number, expected delivery date, and payment receipt. For instance, it could respond, "Your order for red running shoes size 10 has been placed. Your order number is 12345. Expect delivery in 5 days."

Examples and Differentiation from Other Categories

Example in Retail and E-commerce

In retail and e-commerce, chatbots can guide customers through the entire shopping process, from product inquiry to payment, providing a holistic shopping experience. This task goes beyond answering questions or scheduling; it involves transactional interactions and decision facilitation.

Example in Food Services

In food service, chatbots can take orders, suggest menu items based on customer preferences, and handle payment for orders, streamlining the ordering process in a fast-paced environment.

Differentiation

The unique aspect of processing orders and payments by chatbots is the combination of customer service with transaction processing. It involves not just providing information or booking appointments but actively assisting in decision-making and securely handling financial transactions.

m Conclusion

AI chatbots' capability in processing orders and payments demonstrates their advanced role in facilitating complex customer interactions, combining service provision with transactional processing. This functionality is key in retail and e-commerce sectors, where it enhances the customer shopping experience by offering an efficient, interactive, and secure purchasing journey

Providing Personalized Recommendations

Providing personalized recommendations through AI chatbots is a task that exemplifies the advanced capabilities of AI in understanding and responding to individual customer preferences. This function sets itself apart from more straightforward tasks like FAQs or appointment scheduling by involving a more complex level of data analysis and personalization.

How AI Chatbots Provide Personalized Recommendations

Data Analysis and Learning

AI chatbots tasked with providing personalized recommendations are equipped with machine learning algorithms. These algorithms analyze customer data, including past purchases, browsing history, preferences expressed during interactions, and even social media behavior if integrated. This analysis forms the basis for understanding individual customer preferences.

Input: Customer Interactions and Data

Inputs for this task include direct customer queries or interactions, and data points collected from various customer touchpoints. For example, a customer on a retail site might ask the chatbot, "What are some good book recommendations for me?" The chatbot then uses available data, such as the customer's previous purchases or viewed items, to formulate its recommendations.

Personalization Process

The chatbot uses the analyzed data to create a customer profile, which helps it predict preferences and suggest relevant products or services. Advanced NLP allows the chatbot to understand and interpret the customer's query in context, further refining the recommendations.

Output: Tailored Recommendations

The output is a set of personalized recommendations, presented to the customer in a conversational manner. For instance, in response to the book query, the chatbot might say, "Based on your interest in historical fiction, you might enjoy 'The Nightingale' by Kristin Hannah."

Examples and Unique Aspect

Example in Retail

In an online retail setting, a chatbot can suggest clothing items or accessories to a customer based on their past shopping behavior, size preferences, and style choices, offering a curated shopping experience.

Example in Entertainment

For a streaming service, a chatbot can recommend movies or TV shows based on the customer's viewing history and ratings they've given to other content, enhancing the user experience with personalized entertainment options.

Distinctive Characteristic

The distinctive aspect of providing personalized recommendations lies in the chatbot's ability to interpret and utilize complex customer data to make predictive suggestions. This involves not just responding to queries but actively engaging in customer preference analysis and delivering tailored responses, making it a sophisticated application of AI in enhancing customer experiences.

Conclusion

The task of providing personalized recommendations highlights the advanced and interactive nature of AI chatbots, where they act as intelligent assistants, enhancing customer experiences with tailored suggestions. This capability is crucial in sectors like retail and entertainment, where personalization is key to customer engagement and satisfaction.

Conducting Surveys and Collecting Feedback

Conducting Surveys and Collecting Feedback through AI chatbots is a task that leverages the conversational capabilities of chatbots to engage users and gather their opinions and experiences. This functionality is distinct in its approach, as it actively seeks out user input for insights rather than just responding to user-initiated queries or facilitating transactions.

How AI Chatbots Conduct Surveys and Collect Feedback

m Setup and Integration

To perform this task, chatbots are programmed with specific survey questions or feedback prompts. They are integrated into customer interaction points, such as after a purchase, post-service interaction, or at regular intervals within an app or website.

Input: User Engagement

The input for this task involves initiating a conversation with users to solicit their feedback. For example, after a customer service interaction, the chatbot might ask, "Would you like to rate your experience today?" or "Can you provide feedback on our service?"

Interactive Data Collection

The chatbot guides users through the survey or feedback process in an interactive manner. It presents questions, records responses, and can even probe deeper based on previous answers. This interactivity, often conversational in nature, is key in encouraging user participation and obtaining meaningful data.

Output: Compiled Feedback and Analytics

The output from this task is the aggregated feedback and survey data, which is then analyzed to derive insights. Chatbots can compile data in formats conducive to analysis, such as numerical ratings, categorized comments, or trend identification.

Examples and Unique Aspect

Example in Market Research

In market research, a chatbot can collect consumer opinions on a new product, asking questions about preferences, price points, and perceived value. The chatbot can adapt its questions based on the initial responses to gather more nuanced data.

Example in Customer Service

Post-interaction surveys in customer service, where the chatbot asks customers to rate their satisfaction and provide comments on the service received, offer critical insights into performance and areas for improvement.

Distinctive Characteristic

The distinctive characteristic of conducting surveys and collecting feedback is the chatbot's role as an active data collector. Unlike other tasks focused on providing information or services to users, this function is about extracting information from users, making it crucial for businesses seeking customer insights and data-driven decision-making.

m Conclusion

The ability of AI chatbots to conduct surveys and collect feedback illustrates their role in interactive data collection and analytics. This functionality is invaluable across sectors for understanding customer preferences, gauging satisfaction, and informing business strategies. The engagement-driven nature of this task showcases the chatbots' utility in not just serving but also learning from the user base

Language Translation and Multilingual Support

Language Translation and Multilingual Support through AI chatbots represent a significant leap in making digital interactions more inclusive and globally accessible. Unlike tasks centered around a specific operational function or customer interaction in a single language, this capability extends the utility of chatbots across linguistic barriers, making them versatile tools in a globalized world.

How AI Chatbots Perform Language Translation and Offer Multilingual Support

Advanced Language Processing Capabilities

For a chatbot to offer real-time translation and multilingual support, it must be equipped with advanced Natural Language Processing (NLP) capabilities. These NLP tools are not only designed to understand and interpret different languages but also to capture the nuances, dialects, and cultural contexts within those languages.

Input: Multilingual User Queries

The input in this scenario involves queries or commands from users in different languages. For instance, a customer might ask a question in Spanish on an English-based customer service platform: "¿Cuánto cuesta este producto?"

Translation and Response Mechanism

Upon receiving a query in a foreign language, the chatbot employs real-time translation algorithms to interpret the question and convert it into the primary language of the system. It then processes the query, formulates a response, and translates that response back into the user's language. In our example, the chatbot would translate the question to English, find the price of the product, and reply in Spanish.

Output: Accurate and Contextual Responses in User's Language

The output is a coherent, contextually appropriate response in the user's language, ensuring effective communication. For example, the chatbot might reply, "El precio del producto es de 50 dólares."

Examples and Unique Aspect

m Example in Travel and Hospitality

For a hotel booking chatbot, multilingual support means assisting customers from around the world in their native languages, from room booking inquiries to requests for local information, enhancing the user experience for international travelers.

Example in International Customer Service

In customer service, a chatbot capable of handling multiple languages can assist customers from different countries, breaking down language barriers and expanding the reach of the business.

Distinctive Characteristic

The key characteristic that sets language translation and multilingual support apart is its focus on breaking down language barriers and enhancing accessibility. This functionality expands the reach and applicability of chatbots to a global audience, unlike other tasks that are generally language-specific or operationally focused.

Conclusion

Language translation and multilingual support capabilities of AI chatbots underscore their role in fostering global communication and inclusivity. By bridging language gaps, these chatbots not only improve customer experiences but also empower businesses to operate on an international scale, making them indispensable tools in our increasingly connected world.

Lead Generation and Qualification

Lead Generation and Qualification via AI chatbots is a task that uniquely blends the art of conversation with strategic data collection, setting it apart from other chatbot functions like simple information provision or transaction processing. This capability is vital in identifying potential customers and assessing their interest or needs, which is critical for the sales funnel in various industries.

How AI Chatbots Perform Lead Generation and Qualification

Strategic Conversation Initiation

For lead generation and qualification, chatbots are programmed to initiate and steer conversations in a way that naturally uncovers potential customer interests or needs. Unlike other tasks where the chatbot responds to direct queries, here the chatbot takes a more proactive role in the conversation.

Input: User Engagement and Responses

The process begins with the chatbot engaging a website visitor or a user on a digital platform. It might start with a greeting, followed by questions or prompts designed to gauge the user's interest. For example, a real estate chatbot might ask, "Are you looking for a new home?" or "What features are important to you in a property?"

Interactive Qualification Process

Based on the user's responses, the chatbot further tailors its questions to qualify the lead. This involves understanding the user's specific requirements, budget, timelines, or preferences. The chatbot uses NLP to interpret the user's responses accurately and decides the next best questions to ask, guiding the conversation effectively.

Output: Qualified Leads and Insights

The output from this interaction is a set of qualified leads with details about their specific interests and requirements. For instance, the chatbot might conclude that a user is interested in a two-bedroom apartment in a specific area and within a certain price range. This information is then passed on to human sales agents for follow-up.

Examples and Unique Aspect

Example in Sales

In a retail setting, a chatbot on an e-commerce site can interact with shoppers, asking about their preferences or what they are looking to buy, effectively identifying potential leads for targeted products or promotions.

Example in B2B Services

For B2B services, chatbots on corporate websites can qualify leads by asking visitors about their business needs, the size of their company, or their decision-making role, thus aiding the sales team in prioritizing leads.

Distinctive Characteristic

Lead generation and qualification by chatbots differ from other tasks due to their proactive, conversational approach aimed at identifying and qualifying potential customers' interests, a key step in the sales and marketing process.

Conclusion

The task of lead generation and qualification showcases the advanced capability of AI chatbots to not only interact but also strategically guide conversations for business development purposes. This function is crucial for sales, real estate, and B2B services, where identifying and understanding potential customer needs are fundamental to business growth

Providing Basic Technical Support

Providing Basic Technical Support through AI chatbots involves a unique combination of guiding users through problem-solving processes and imparting specific technical knowledge. This task differentiates itself from other chatbot applications like information retrieval or appointment scheduling by requiring the chatbot to understand and respond to technical issues with precise, step-by-step instructions.

How AI Chatbots Provide Basic Technical Support

Advanced Problem-Solving Logic

AI chatbots designed for technical support are programmed with a knowledge base of common technical issues and their resolutions. They use a logic-based approach to diagnose problems based on user input and guide them through troubleshooting steps.

Input: User-Reported Technical Issues

The input for this task involves users reporting a technical issue they are facing. For instance, a user might approach a chatbot with a problem like, “My internet connection is not working” or “I can’t access my email account.”

Diagnostic and Guided Support Process

Upon receiving a query, the chatbot uses NLP to understand the nature of the issue. It then asks follow-up questions to diagnose the problem accurately. For example, it might ask, “Are you connected to Wi-Fi?” or “Are you receiving any error messages?” Based on the responses, the chatbot identifies the issue and provides a step-by-step guide to resolve it.

Output: Resolved Issues and User Instructions

The output is a series of instructions or actions that the user can follow to resolve their technical issue. In our examples, the chatbot might guide the user to check their network settings or reset their password, providing clear and concise directions.

Examples and Unique Aspect

Example in IT Support

In an office setting, an AI chatbot can assist employees with common IT problems, such as printer malfunctions or software errors, reducing the workload on the IT department.

Example in Telecommunications

For a telecom company, chatbots can help customers troubleshoot issues with their mobile devices or service interruptions, offering quick solutions without the need for a service call.

Distinctive Characteristic

The distinctive aspect of providing basic technical support is the chatbot's ability to interactively diagnose and resolve specific technical issues. Unlike tasks focused on providing general information or conducting transactions, this function requires an understanding of technical problems and the ability to guide users through detailed resolution processes.

Conclusion

The task of providing basic technical support showcases AI chatbots' capability to function as interactive, problem-solving tools. This application is particularly valuable in IT and telecommunications, where immediate, accurate technical assistance can significantly enhance user experience and operational efficiency

Conclusion

As we conclude this insightful unit on the diverse capabilities of AI chatbots, it's clear that these advanced tools are far more than mere digital assistants for answering queries. We've explored a broad spectrum of tasks that AI chatbots are equipped to handle, each with its unique complexities and applications across various industries.

From responding to FAQs with promptness and precision to managing complex tasks like appointment scheduling and order processing, AI chatbots have demonstrated an extraordinary capacity to enhance efficiency and improve user experiences. We delved into how these chatbots provide personalized recommendations, tapping into advanced data analysis to tailor responses to individual user preferences. This level of personalization is a testament to the sophisticated algorithms and machine learning capabilities that drive modern chatbots.

Furthermore, we observed how AI chatbots are revolutionizing the field of customer service and support, offering step-by-step guidance for technical troubleshooting, a task that once demanded significant human intervention. Their ability to conduct surveys and gather feedback marks them as invaluable tools for businesses in garnering insights and understanding customer needs and preferences.

In understanding these various functionalities, it's evident that AI chatbots stand apart in their ability to automate and streamline a multitude of tasks, ranging from the simple to the complex. Their versatility in handling different types of tasks, each requiring a unique approach and set of capabilities, highlights the expansive role they play in today's digital landscape.

Moreover, the exploration of language translation and multilingual support underscored the global reach and inclusivity that chatbots bring to digital interactions, breaking down language barriers and enhancing communication.

The examples provided throughout the unit not only illustrated the practical applications of these capabilities but also shed light on the transformative impact AI chatbots have on business processes. From increasing operational efficiency to enhancing customer satisfaction and driving sales, the benefits of chatbots are multi-dimensional.

In summary, this unit has equipped us with a comprehensive understanding of the myriad tasks AI chatbots can perform, each contributing uniquely to the efficiency, effectiveness, and evolution of business operations. As we advance, it's clear that AI chatbots will continue to be integral players in the digital era, driving innovation and shaping the future of customer and business interaction.

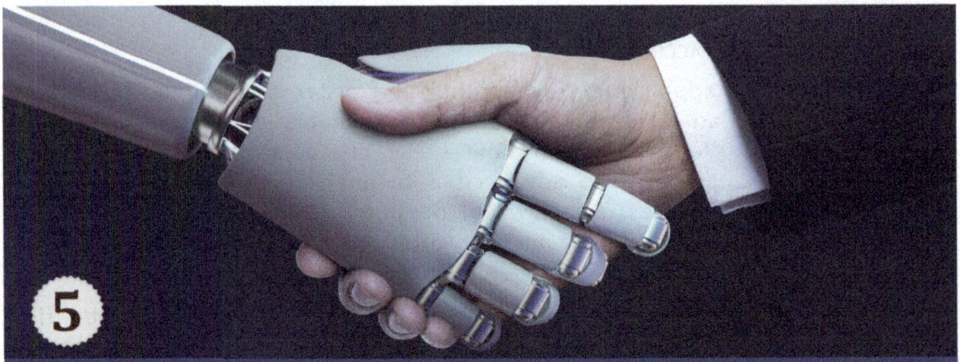

5

CASE STUDIES OF NOT FOR PROFIT'S AND GOVT. AGENCIES USING AI CHATBOTS

Learning Outcomes:

By the end of this unit, learners should be able to:

- Understand the practical application of AI chatbots in various Not For Profit's and Govt. Agencies through detailed case studies.
- Gain insights from real-world examples of AI chatbot implementations, their benefit and impact on operations.
- Apply these insights to evaluate the potential impact of AI chatbots on their own business

Welcome to Chapter 5, where we venture into the transformative world of AI chatbots within the Not For Profit sector and Governmental Agencies. Through a series of detailed case studies, we will explore the practical applications, benefits, and operational impacts of AI chatbots in this unique and vital sector.

This sector encompasses a wide range of organizations, including charities, educational institutions, health care providers, and social welfare bodies, each with their own set of challenges and objectives. In this context, AI chatbots emerge as innovative solutions, offering cost-effective, efficient, and scalable means to enhance service delivery, engage with stakeholders, and optimize operations.

Understanding the Practical Application in Not For Profits

AI chatbots, with their ability to automate interactions, provide information, and guide users, have found a special place in this Sector. These organizations often operate under budget constraints and face the challenge of meeting high service demands with limited resources. AI chatbots step in as a game-changer, offering a way to extend reach and capabilities without proportionally increasing costs.

Insights from Real-World Implementations

Each case study in this chapter is carefully selected to represent a spectrum of applications and the diverse needs of organizations in this sector. We will delve into examples where AI chatbots have been implemented for purposes like fundraising, community engagement, educational support, health advice, and more. These real-world examples will not only illustrate the functional capabilities of AI chatbots but also highlight the strategic thinking behind their integration.

Evaluating Impact and Benefits

The true value of AI chatbots lies in their ability to make a real difference - be it through increased donor engagement, improved service delivery, or enhanced community outreach. We will examine how these organizations have leveraged AI chatbots to overcome operational challenges, enhance efficiency, and, most importantly, better serve their causes and communities.

For learners looking to evaluate the potential of AI chatbots and their application this chapter will serve as a comprehensive guide. The insights and learnings from these case studies will provide a valuable perspective on how AI technology can be adapted and implemented to meet specific organizational needs and goals.

As we embark on this journey through Chapter 5, prepare to be inspired by the innovative ways in which organizations in this sector are harnessing the power of AI chatbots. Their stories are not just about technology adoption but about creating impact, driving change, and enhancing the quality of service in the face of challenges

Amarillo's "Digital Human" - Bridging Language Barriers in Public Services

Introduction to Amarillo's Innovative Initiative

The city of Amarillo, Texas, renowned for its diverse population and significant refugee community, embarked on an ambitious project to enhance its public service accessibility. Recognizing the need to cater to a multilingual populace, the city has been developing a "digital human" - an advanced AI chatbot with multilingual capabilities set to revolutionize interactions between the city and its residents.

The Need for Multilingual AI Chatbot Integration

Amarillo's unique demographic, with a middle school alone hosting students speaking 62 languages and dialects, presented a formidable challenge in providing equitable public services. The city's initiative, driven by the aim to offer equal service to all citizens, especially refugees, led to the adoption of AI technology to overcome language barriers efficiently.

How the "Digital Human" Works

- Advanced AI and Language Processing: Amarillo's digital assistant, powered by conversational AI and integrated with NLP (Natural Language Processing) capabilities, can understand and communicate in 62 different languages. This technology allows for real-time translation and personalized interaction with users.

- Input and Interaction: Users can interact with the digital assistant through the city's website, asking various questions ranging from simple inquiries like library hours to more complex requests such as booking spots in local parks.
- Output and Service Provision: The digital human provides immediate, accurate responses and directs users to appropriate web pages for further actions. It can walk residents through processes like booking public spaces or accessing city services, all in the user's preferred language.

The Impact and Benefits

- Enhanced Accessibility and Inclusivity: By offering support in multiple languages, the digital assistant ensures that all residents, irrespective of their linguistic background, can access city services efficiently.
- Cost-Effective Solution: This AI-driven approach circumvents the need for a large team of interpreters, offering a cost-effective solution to the language diversity challenge.
- Improving Citizen-Government Interaction: The tool transforms traditional interactions into seamless, conversational exchanges, enhancing the user experience and fostering greater engagement with municipal services.

Collaboration and Technology

Amarillo partnered with UneeQ, specializing in digital humans, and Dell Technologies to develop this sophisticated AI tool. This collaboration highlights the merging of public service needs with cutting-edge technology.

Broader Implications and Future Outlook

Amarillo's initiative sets a precedent for other cities grappling with similar challenges, showcasing how technology can be leveraged to enhance public service delivery. The digital human project, once fully operational, is expected to cater to 60%-80% of the city's residents, marking a significant step towards creating a more connected and inclusive smart city.

Conclusion

Amarillo's digital human represents a significant leap in public service provision, using AI to bridge language gaps and ensuring equal access to services for a diverse population.

Sophia - Empowering Survivors of Domestic Violence with AI Chatbot Assistance

Introduction to Sophia - A Lifeline for Survivors

Sophia is an innovative AI chatbot designed to assist survivors of domestic violence. Operating globally, Sophia offers confidential support 24/7, transcending geographical and language barriers. Its unique approach lies in the "G.A.L." system - Gather evidence, Assess rights, and Learn about options, making it an exemplary case study under the category of "Providing Personalized Recommendations."

How Sophia Works: The G.A.L. System

1. Gather Potential Evidence: Sophia is programmed to help survivors discreetly gather and document evidence of abuse. This function involves guiding users through safely recording incidents, which can be crucial for legal proceedings or seeking help. This step goes beyond typical data retrieval, as it involves sensitive and strategic information gathering.

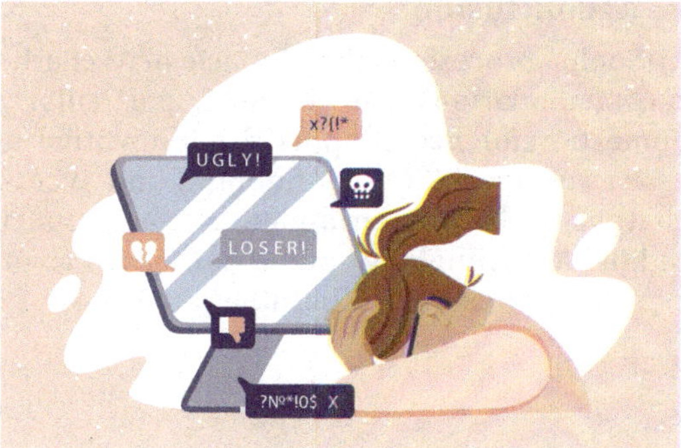

2. Assess Your Rights: In this phase, Sophia provides users with information about their

legal rights, specific to their location and situation. This personalized legal guidance is crucial for survivors who may not be aware of their rights or the legal resources available to them.

3. Learn About Your Options: Sophia outlines various options available for survivors, including seeking shelter, legal action, and contacting support services. This aspect of the chatbot's functionality is akin to providing personalized recommendations, tailored to the specific needs and circumstances of the user.

Application and Impact

- Global Reach and Accessibility: Given its digital nature, Sophia is accessible to anyone with internet access, offering support to survivors across the world, regardless of their location.
- Confidentiality and Safety: Sophia's design prioritizes user safety and confidentiality, which is critical for survivors of domestic violence. This level of privacy assurance distinguishes Sophia from other chatbots in more general domains.
- Multilingual Support: To cater to a diverse user base, Sophia supports multiple languages, ensuring that language barriers do not impede access to vital information and support.

Real-World Example

Consider a scenario where a survivor in a remote area, unsure of their legal rights and available options, reaches out to Sophia. The chatbot discreetly guides them through documenting evidence, informs them of their rights, and provides options for immediate and long-term safety. This guidance could potentially be life-saving and certainly empowering for the survivor.

Conclusion

Sophia represents the pinnacle of AI chatbots' potential in providing personalized recommendations and support, particularly in sensitive and critical areas like domestic violence support. Its innovative use of AI technology to empower and assist survivors in a confidential and accessible manner sets a new standard for chatbots in humanitarian and social support sectors. Sophia's case is a testament to how AI technology can be harnessed to make a real difference in people's lives, especially those in vulnerable situations

Clara-Revolutionizing Blood Donation with AI Chatbot Technology

Introduction to Clara - The American Red Cross's AI Innovation

Clara is an AI-powered chatbot developed by the American Red Cross, designed to revolutionize the way individuals engage with blood donation processes. Named after Clara Barton, the esteemed founder of the American Red Cross, Clara the chatbot embodies the spirit of innovation and compassion in public health services.

Role and Functionality of Clara

Clara is integrated into the American Red Cross's website, redcrossblood.org, serving as an intelligent assistant to potential and existing blood donors. Utilizing the power of AI, particularly Natural Language Processing (NLP), Clara is equipped to handle a variety of tasks and queries related to blood donation.

1. Appointment Scheduling and Management: Clara simplifies the process of scheduling and managing appointments for blood, Power Red, and platelet donations. Users can interact with Clara to find suitable times for their donations and manage their appointments, enhancing the efficiency and convenience of the donation process.

2. Providing Information on Blood Donation: Clara is programmed to answer a wide range of questions about blood donation. This includes details about the donation process, the impact of donations, and other ways to contribute, such as sponsoring a drive or making financial donations.

3. Eligibility Queries: Addressing one of the most crucial aspects of blood donation, Clara assists users with eligibility questions. This encompasses queries about travel, medications, and lifestyle factors that might affect a person's eligibility to donate blood.

Clara's Technological Backbone

- Natural Language Processing: Clara's ability to process and understand natural language data enables her to interpret and respond accurately to user queries. This technology is key to making Clara user-friendly and accessible to a broad audience.

Clara's Extension on RedCross.org

- Disaster Response Information: Beyond blood donation, Clara also provides essential information on RedCross.org in times of disaster, aiding both those in need and those looking to help. This dual functionality showcases Clara's versatility and the Red Cross's commitment to leveraging technology in humanitarian efforts.

Clara's Impact and Innovation

- Enhanced Donor Experience: Clara significantly improves the donor experience by simplifying appointment scheduling and providing instant answers to donation-related questions.
- Educational Tool: By educating users about the blood donation process and eligibility, Clara plays a vital role in demystifying misconceptions and encouraging more people to donate.
- Emerging Technologies: Clara's use of advanced AI and NLP demonstrates the American Red Cross's commitment to innovation, aligning with Clara Barton's legacy of pioneering in public health and humanitarian services.

Conclusion

Clara represents a significant stride in the application of AI technology in the Not For Profit sector, particularly in public health and emergency response. By offering an intuitive and informative platform for blood donors, Clara not only streamlines the donation process but also perpetuates the legacy of Clara Barton in using innovative solutions to save lives and assist those in need

Case Study: MISSI – Official Mississippi State Chatbot

Introduction: MISSI is an AI-powered chatbot developed for the state of Mississippi, designed to provide residents with easy access to crucial state information.

Functionality: The chatbot offers information on a variety of topics including taxes, transit, tourism, and elder care. It is specifically tailored to provide conversational responses and detailed local knowledge.

Impact and User Satisfaction: State officials have reported extremely high user satisfaction with MISSI, attributed to its user-friendly conversational abilities and comprehensive knowledge of local state information.

Conclusion: MISSI exemplifies the successful integration of AI in state governance. By providing residents with instant access to important information, it enhances public service delivery and demonstrates the potential of AI chatbots in improving citizen-government interaction.

Roo by Planned Parenthood - An AI Chatbot for Teen Sexual Health Education

Introduction to Roo: Addressing the Delicate Needs of Teens

In a world where teens are increasingly turning to the internet for information, Planned Parenthood identified a critical gap in sexual health education. Recognizing that teens often hesitate to discuss sensitive topics like sex with parents or on social media platforms due to privacy concerns, Planned Parenthood launched Roo, an AI chatbot designed to provide reliable and confidential information on sexual health.

The Challenge: Providing Accurate Information in a Safe Space

The challenge was to create a digital solution that offered a safe, private, and judgment-free environment for teens to learn about sexual health. Traditional online platforms like Google or social media were not viable options for teens due to privacy concerns and the prevalence of misinformation.

How Roo Works

- Text Messaging Interface: Roo was designed to mimic a text messaging experience, creating a familiar and comfortable interaction mode for teens. This approach made the chatbot approachable and user-friendly.
- Anonymous Interaction: Understanding the sensitivity of the subject, Roo allows users to ask questions anonymously, ensuring privacy and encouraging open communication.
- AI-Powered Responses: Utilizing AI and Natural Language Processing, Roo can answer a wide range of questions related to sex and sexual health. From general topics like birth control and pregnancy to more specific queries about masturbation, Roo provides accurate and medically sound advice.
- Web-Based Accessibility: Research showed that teens were reluctant to download another app, leading to Roo being made accessible directly via the web. This decision enhanced accessibility and user convenience.

The Impact of Roo

- Bridging the Education Gap: In states with varying sex education policies and rampant online misinformation, Roo serves as a reliable source of factual information.

- Empowering Teens: By providing accurate information in a private setting, Roo empowers teens to make informed decisions about their sexual health.
- Innovative Use of Technology: Roo's implementation showcases an innovative use of AI in public health, specifically tailored to the needs and preferences of its target audience - teens.

Conclusion

Roo stands as a testament to the potential of AI chatbots in addressing sensitive and critical health education needs. Planned Parenthood's initiative reflects a deep understanding of the target demographic's preferences and concerns. Roo's success highlights how AI technology can be leveraged to create safe, private, and effective educational platforms, especially in areas where traditional methods of education may fall short. This case study demonstrates the power of AI in bridging knowledge gaps and addressing public health challenges in a digital age

Resistbot - Empowering Civic Engagement Through AI Chatbot Technology

Introduction to Resistbot: A Tool for Democratic Participation

In an era where civic engagement is more crucial than ever, Resistbot emerges as a pioneering solution in the United States to facilitate communication between citizens and their elected officials. This AI-powered chatbot service simplifies the process of political participation, making it easy and accessible for people to exercise their democratic rights from their mobile phones.

What Resistbot Offers: Features and Functionalities

- Letter Composition and Delivery: One of the primary functions of Resistbot is to enable users to compose and send letters to local, state, and federal representatives in under two minutes. This feature demystifies and streamlines the traditionally complex process of contacting government officials.
- Multi-Platform Compatibility: Operable through various messaging apps like Apple Messages, WhatsApp, Messenger, Instagram, Telegram, and SMS, Resistbot ensures accessibility across different digital platforms.
- Automatic Official Identification: By automatically identifying elected officials based on the user's location, Resistbot removes the barrier of

needing to know specific contact details.

- Public Letter Sharing (Optional): Users have the option to publish their letters online or send them to local newspapers, enhancing the transparency and reach of their messages.

Additional Features Enhancing Civic Participation

- Voter Registration Confirmation: The service helps confirm voter registration status, an essential feature in fostering active electoral participation.
- Locating Polling Places and Town Halls: Resistbot provides information on local polling places and town hall events, encouraging in-person civic nvolvement.
- Volunteer Opportunities and Voting Information: The chatbot assists users in finding volunteer opportunities and accessing crucial voting information, further bolstering civic engagement.

Benefits of Resistbot

- Convenience and Accessibility: The ease of use on mobile phones makes Resistbot a convenient tool for busy or remotely located citizens.
- Low Barrier to Political Participation: By simplifying the process, Resistbot encourages even those who are not typically politically active to engage with their representatives.
- Amplification of Individual Voices: The collective action campaigns facilitated by Resistbot amplify individual efforts, creating a more significant impact.

Considerations and Non-Partisan Nature

- Non-Partisan Platform: Resistbot serves all users, regardless of their political beliefs, underscoring its commitment to democratic participation.
- Non-Profit Operation: As a non-profit, Resistbot relies on donations, highlighting its community-driven nature.

Conclusion

Resistbot stands as a prime example of how AI chatbots can be leveraged beyond conventional commercial or informational uses, venturing into the realm of civic technology. By providing a straightforward, efficient, and inclusive platform,

Resistbot significantly contributes to increasing civic engagement, making it easier for citizens to be heard and participate in the democratic process. This case study illustrates the transformative potential of AI in empowering individuals and enhancing democratic participation in the modern digital age

Case Study: RAMMAS – Dubai Electricity and Water Authority's Chatbot

Introduction: RAMMAS is an AI-powered chatbot introduced by the Dubai Electricity and Water Authority (DEWA) to enhance customer service and streamline operations.

Implementation and Functionality: Launched in 2017, RAMMAS serves as a 24/7 customer service agent. It enables customers to pay bills, report issues, and access account information with ease. This innovative tool was designed to handle a high volume of inquiries, efficiently managing over 700,000 inquiries in its first year alone.

Impact and Benefits: The implementation of RAMMAS significantly reduced the workload on human customer service representatives, effectively managing a volume of inquiries that would have been unmanageable otherwise. It provided customers with quick and reliable access to essential services, thereby improving customer satisfaction and operational efficiency.

Conclusion: RAMMAS stands as a testament to the effective use of AI in government services, showcasing how technology can revolutionize customer interaction and service delivery in public utilities. Its success in managing large volumes of inquiries and improving customer service efficiency exemplifies the transformative potential of AI in the public sector.

> *Think of AI chatbots as friendly bridges between citizens and their governments. They break down information walls, open doors to easier services, and build a two-way street for feedback and insights.*

Crisis Text Line - Enhancing Crisis Support with AI Technology

Introduction to Crisis Text Line and Its Use of AI

Crisis Text Line, a not-for-profit organization, has revolutionized the way crisis intervention is provided through text messaging. While the core of its service relies on trained human crisis counselors, AI technology plays a significant supporting role in enhancing the effectiveness and efficiency of their services.

AI in Triage and Prioritization

- AVA (Artificial Intelligence Virtual Assistant): Crisis Text Line utilizes AVA, an AI algorithm designed to analyze incoming text messages. Its primary function is to quickly identify texters who are at the highest risk of suicide or self-harm.
- Triage System: Once high-risk cases are flagged by AVA, they are prioritized for immediate attention by human counselors. This ensures that the most vulnerable texters receive the urgent support they need without unnecessary delay.

Resource Identification Through AI

- Assisting Counselors with Resources: AI tools at Crisis Text Line are used to aid counselors in rapidly locating and sharing relevant resources with texters. These resources range from suicide prevention hotlines to mental health websites and local crisis centers, tailored to the specific needs of each texter.

Leveraging AI for Data Insights and Research

- Conversation Data Analysis: The organization employs AI to sift through vast amounts of conversation data. This analysis helps identify patterns, trends, and risk factors in the conversations.
- Service Improvement: Insights derived from AI analysis are crucial for refining counseling strategies, enhancing training programs, and developing new resources to better serve those in crisis.

Piloting AI Chatbots for Specific Needs

- Tessa - AI Chatbot for Eating Disorders: Crisis Text Line is piloting an AI chatbot named Tessa, specifically aimed at providing support for individuals struggling with eating disorders. This pilot project explores the potential of AI chatbots in offering initial support and triage.
- Balancing AI and Human Engagement: The organization is keen on exploring the capabilities of AI chatbots while ensuring that human counselors are always available for direct and empathetic engagement.

Ethical Considerations in AI Usage

- Maintaining Human Connection: Crisis Text Line emphasizes the irreplaceable value of human connection and empathy in crisis intervention a facet that AI cannot replicate.

- Commitment to Ethical AI Use: The organization upholds a strong commitment to ethical AI usage, maintaining transparency in its AI operations and prioritizing the privacy and safety of its users.

Conclusion

Crisis Text Line's application of AI technology exemplifies how AI can play a vital supporting role in enhancing the efficiency and effectiveness of crisis intervention services. While AI aids in triage, resource allocation, and data analysis the organization maintains a careful balance, ensuring that the human element remains central to its mission. This case study demonstrates a thoughtful and innovative approach to integrating AI in mental health support services, setting a benchmark for ethical and effective use of technology in the Not For Profit sector and Govt. Agencies

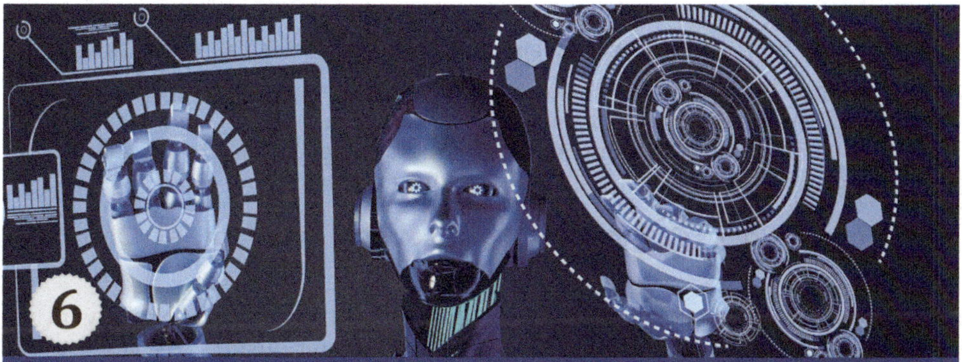

PRACTICAL IMPLEMENTATION OF AI CHATBOTS

Learning Outcomes:

By the end of this unit, learners should be able to:

- Understand the steps involved in implementing AI chatbots in a business.
- Recognize the considerations necessary for effective AI chatbot implementation.
- Understand how to measure the impact and success of AI chatbot implementation.
- Apply this practical knowledge to implement AI chatbots in their own businesses and measure their impact.

Welcome to Chapter 6, a pivotal section where we get down to the brass tack, the nitty gritty of beginning to benefit from what we have learnt so far. Having explored the essence, evolution, and diverse applications of AI chatbots across various sectors, including their significant role in Not For Profit organizations, we now turn our attention to the practicalities of implementing AI chatbots within your organization. This chapter is dedicated to guiding you through the process of identifying, selecting, and deploying the right AI chatbot solution to address specific needs and enhance your operational efficiency.

Identifying Processes for AI Chatbot Implementation

The first step in implementing an AI chatbot is identifying the processes within your organization that can benefit from automation and enhanced interaction. We will delve into how to analyze your current operations to pinpoint areas where AI chatbots can bring the most value. Whether it's improving customer service, streamlining internal communications, or automating routine tasks, identifying the right processes is crucial for a successful implementation.

Conducting a Comprehensive Process Audit for AI Chatbot Implementation

The first and foremost step in implementing AI chatbots effectively in an organization is to conduct a comprehensive audit of your business processes. This audit aims to identify key areas where chatbots can be most beneficial. Here's how you can approach this crucial step.

Identifying Key Business Operations

- List All Major Operations: Start by creating an exhaustive list of all major departments and operations within your organization. Typical areas might include customer service, sales, human resources (HR), information technology (IT) support, marketing, and finance. For example, a retail company might focus on operations like online customer support, inventory management, and order processing.
- Understand Each Department's Role: Delve into the specifics of what each department does and how it contributes to your overall business goals. In a healthcare setting, for instance, key operations might include patient scheduling, medical inquiries, and patient follow-up procedures.

Evaluating Current Performance Metrics

- Assess Performance Indicators: For each department, assess performance using relevant metrics. In customer service, look at metrics like average response time to customer queries, first contact resolution rate, and overall customer satisfaction scores. For sales, evaluate lead response time, conversion rate, and sales cycle length.
- Collect and Analyze Data: Gather data that reflects the current performance of each operation. Tools like CRM systems, customer feedback surveys, and performance reports can be invaluable. For instance, an IT support department might analyze ticket resolution times and the volume of repeat queries.

Practical Examples and Tips

- Example in Customer Service: A telecom company might find that its customer service team is overwhelmed with inquiries about basic account information, leading to long wait times. By implementing a chatbot, these routine inquiries can be automated, allowing human agents to focus on more complex issues.
- Tip for Sales Department: In sales, if the data shows a high volume of online queries about product specifications which delays the sales process, implementing a chatbot for pre-sales inquiries can streamline operations and improve conversion rates.
- Example in HR: For the HR department, high volumes of repetitive queries about company policies or employee benefits could indicate a need for an HR chatbot to provide instant responses, thereby reducing employee frustration and HR workload.

Additional Considerations

- Employee Feedback: Conduct surveys or interviews with employees to understand the challenges they face in their day-to-day operations. Employee insights can be particularly revealing in identifying inefficiencies.
- Customer Journey Mapping: For customer-facing operations, map out the customer journey to identify pain points. Understanding where customers face difficulties can guide you in placing chatbots effectively.
- Ongoing Evaluation: Remember, a process audit is not a one-time activity.

Continuously monitor and evaluate the performance metrics post-chatbot implementation to gauge effectiveness and identify areas for further improvement.

Conclusion

A comprehensive process audit is a critical step in ensuring the successful implementation of AI chatbots in an organization. By methodically evaluating key operations and their performance metrics, you can identify the most impactful areas for chatbot integration.

Identifying Pain Points and Challenges for AI Chatbot Implementation

Once you've audited your business operations, the next crucial step in implementing AI chatbots is identifying pain points and challenges within these operations. This step involves pinpointing inefficiencies and bottlenecks and gathering feedback from both employees and customers. Here's how to approach this process effectively.

Pinpointing Inefficiencies and Bottlenecks

- Examine Operational Flow: Analyze the workflow of each operation to identify where delays and inefficiencies occur. For example, in a customer service department, long hold times or delayed email responses might indicate a bottleneck.
- Identify Repetitive Tasks: Look for tasks that are repetitive and time-consuming. These are often ideal candidates for automation. In an HR department, this could be answering frequent queries about company policies or leave entitlements.
- Assess Resource Allocation: Determine if there are areas where staff are overwhelmed with work, particularly with tasks that do not require human judgment. For instance, a sales team might be spending an excessive amount of time qualifying leads instead of closing sales.

Gathering Feedback

- Employee Feedback: Employees are often the first to notice operational inefficiencies. Conduct surveys, interviews, or focus group discussions to gather their insights. Frontline employees, in particular, can provide valuable information about the daily challenges they face.

- Customer Feedback: Use surveys, social media listening, and direct feedback mechanisms to understand customer pain points. In a retail setting, customers might express frustration about the difficulty in getting product information or the lack of immediate assistance.

Practical Examples and Tips

- Example in Customer Service: A retail company might find that customers frequently ask about product availability or shipping details. A chatbot could be implemented to handle these common queries, reducing wait times and improving customer satisfaction.
- Tip for Effective Feedback Gathering: Ensure anonymity in feedback surveys to encourage honest responses. Quantitative data can be complemented with qualitative insights for a more comprehensive understanding.
- Example in IT Support: An IT department may face a high volume of basic technical queries, like password resets, that can easily be automated. Identifying such repetitive tasks can significantly reduce the workload on IT staff.

Additional Considerations

- Quantitative and Qualitative Analysis: Combine data analysis with anecdotal evidence. While metrics provide a quantitative measure of inefficiencies, personal accounts offer qualitative insights into the underlying causes.
- Cross-Departmental Review: Encourage inter-departmental communication during the review process. Challenges in one department may have cascading effects on another, highlighting broader organizational issues.
- Customer Journey Mapping: For customer-facing operations, mapping the customer journey can help identify pain points from the customer's perspective. This could reveal areas where a chatbot can enhance the customer experience.

> *"Chatbots are the tip of the iceberg. The real revolution is the rise of conversational AI."* *–Rodney Brooks, MIT robotics Professor*

Conclusion

Identifying pain points and challenges is a crucial step in ensuring that AI chatbot implementation effectively addresses the real needs of your organization. By understanding both employee and customer perspectives, you can pinpoint exact areas where chatbots can bring the most value, leading to more strategic and successful implementation.

Prioritizing Processes for Chatbot Implementation

After identifying key business operations and their pain points, the next crucial step is to prioritize these processes for chatbot implementation. This involves assessing the potential impact of chatbots on various operations and balancing quick wins with long-term strategic goals. Let's explore how to effectively prioritize processes for chatbot implementation.

Impact Assessment

- Evaluate Time Savings: Assess which processes will benefit most from the time-saving capabilities of chatbots. For instance, if your customer service department is inundated with routine inquiries, implementing a chatbot can significantly reduce response times and free up human agents for more complex tasks.
- Assess Cost Reductions: Identify areas where chatbots can cut operational costs. In many cases, automating repetitive tasks with chatbots can lead to significant savings on labor costs.
- Improvement in Satisfaction Levels: Consider how chatbot implementation could enhance customer or employee satisfaction. For example, a chatbot in HR for answering routine employee queries about policies and benefits can improve employee experience.

Quick Wins vs. Long-Term Goals

- Identify Quick Wins: Look for areas where chatbots can be implemented quickly and yield immediate benefits. These are usually processes with high volumes of routine, rule-based interactions that chatbots can handle efficiently. For example, automating FAQs in customer service can be a quick win, offering immediate relief to the team and improving customer experience.
- Align with Long-Term Strategic Goals: While quick wins are important, it's also crucial to align chatbot implementation with your organization's long-term goals. If one of your strategic objectives is to enhance customer engagement, consider implementing a more advanced chatbot capable of personalized recommendations or lead nurturing.

Practical Examples and Tips

- Example in Sales: If lead qualification is a time-intensive process for your sales team, a chatbot can quickly categorize leads based on predefined criteria, allowing sales representatives to focus on high-potential prospects.

- Tip for Prioritization: Use a scoring system to evaluate and rank processes based on criteria such as potential impact, ease of implementation, and alignment with business goals. This can help in objectively deciding which processes to prioritize.
- Example in Support Services: For an IT support team, a chatbot can handle common queries like password resets or basic troubleshooting, which often make up a large portion of support requests.

Additional Considerations

- Stakeholder Involvement: Involve key stakeholders in the prioritization process. Getting buy-in from different departments ensures a more holistic approach to implementation.
- Pilot Programs: Consider starting with a pilot program in one department or process. This allows you to gauge the effectiveness of the chatbot and make necessary adjustments before wider deployment.
- Continuous Evaluation: Post-implementation, continuously evaluate the impact of the chatbot and be ready to adjust your strategy. As your organization evolves, so too will the opportunities for chatbot applications.

> *"The conversational interface is the natural interface. It's not about typing anymore, it's about talking."* *–Satya Nadella, Microsoft CEO*

Conclusion

Prioritizing processes for chatbot implementation is a critical step that requires a careful assessment of potential impacts, time savings, and cost reductions. Balancing quick wins with long-term strategic goals ensures that your chatbot implementation is not just successful in the short term but also aligns with your organization's future vision. With careful planning and strategic prioritization, chatbots can significantly enhance operational efficiency and customer engagement

Categorizing Needs and Selecting the Right Type of AI Chatbot

Not all chatbots are created equal. In this section, we will explore the different categories of AI chatbots and their unique capabilities. After conducting a comprehensive process audit, identifying pain points and challenges, and prioritizing processes for chatbot implementation, the next crucial step is to assess the suitability of various types of chatbots for your organization's specific needs.

This assessment should align the capabilities of different chatbot types with the identified operational requirements.

From simple FAQ bots to more complex bots capable of handling transactions, appointments, or providing personalized recommendations, understanding the types of chatbots and their functionalities will enable you to choose the one that best suits your organizational needs.

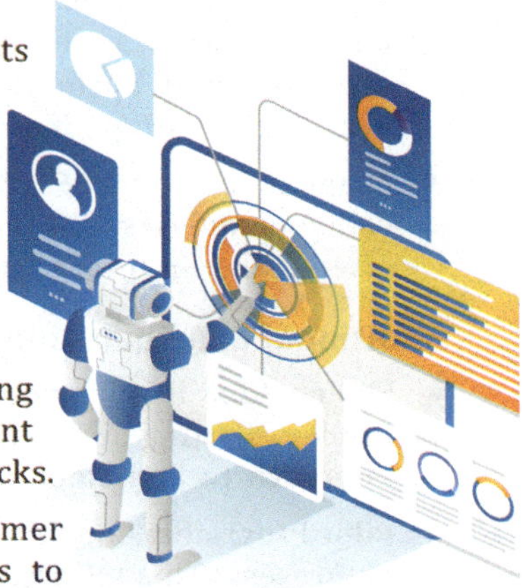

Data Retrieval and Transaction Lookup

- Ideal for processes involving data fetching from databases, like customer account information, order status, or inventory checks.
- Suitable for finance, sales, and customer service departments where quick access to data is essential for operational efficiency.
- Ensure secure integration with your database and compliance with data privacy regulations.

Appointment Scheduling and Management

- Best suited for processes requiring scheduling, rescheduling, or managing appointments.
- Ideal for healthcare, service industries, and internal HR functions.
- Integration with existing calendar and scheduling systems is vital.

How AI Chatbots Process Orders and Payments

- Suitable for e-commerce and retail operations where processing orders and payments are routine.
- Automates the purchasing process and provides a seamless customer experience.
- Security and seamless integration with payment gateways are crucial.

Providing Personalized Recommendations

- Ideal for enhancing customer experience through tailored suggestions based on user behavior and preferences.
- Useful in retail, hospitality, and e-commerce for customized user engagement.
- Requires sophisticated AI algorithms capable of learning from user data.

Conducting Surveys and Collecting Feedback

- Suitable for processes requiring customer or employee feedback.
- Can be implemented across various departments to gather insights and improve services.
- Ensure user-friendly interfaces and anonymity for honest feedback.

Language Translation and Multilingual Support

- Essential for organizations with a diverse customer base or global operations.
- Useful in customer service, tourism, and public service sectors.
- Prioritize accuracy and cultural sensitivity in translation.

Lead Generation and Qualification

- Best for sales and marketing processes where identifying and nurturing leads is crucial.
- Automates the initial stages of the sales funnel, allowing teams to focus on high-value activities.
- AI must be capable of sophisticated interaction to accurately qualify leads.

Providing Basic Technical Support

- Ideal for IT support and troubleshooting standard technical issues.
- Reduces the workload on IT support staff by automating responses to common queries.
- Ensure the chatbot is updated with the latest information and can escalate complex issues to human staff.

> *"The key to successful chatbot implementation is value. When chatbots solve real problems and improve lives, the potential becomes limitless."* *–Demis Hassabis, DeepMind co-founder*

Conclusion

Assessing the suitability of different types of AI chatbots involves aligning their specific capabilities with your organization's identified needs and processes. This alignment ensures that the selected chatbot type not only addresses the immediate operational challenges but also integrates seamlessly into the existing ecosystem, enhancing overall efficiency and user experience.

Step-by-Step Guide to AI Chatbot Implementation

Implementing an AI chatbot in an organization is a multi-faceted process that requires careful planning, selection of the right technology partners, and seamless integration into existing systems. Here's a detailed roadmap to guide you through each step of this journey.

Choosing the Right Chatbot Platform and Partner

In the realm of AI chatbot implementation, selecting the right platform is a critical decision that can significantly influence the success of your project. This decision involves thorough research, evaluation, and testing of potential platforms. Here's a detailed guide on how to choose the right chatbot platform, enriched with real-life examples.

Research and Evaluate Vendors

- Alignment with Business Objectives: Start by ensuring the platform's capabilities align with your specific goals. For instance, Domino's Pizza opted for a chatbot platform that could integrate seamlessly with their ordering system, allowing customers to place orders via a chat interface.
- Evaluation Criteria: Assess platforms based on:
- Features: Look for essential features like NLP capabilities, integration options, response accuracy, and the ability to handle complex dialogues.
- Scalability: The platform should be able to scale as your business grows. For example, a small startup might initially need a basic chatbot but should choose a platform that can handle increased interactions as the business expands.
- Customization Options: Ensure the platform allows customization to meet your specific needs. When U-Report, a global platform designed to engage young people in policy-making, chose a chatbot platform, customization was key to cater to their diverse audience.
- Cost: Consider both initial setup costs and long-term operational costs.

Test Drive Platforms

- Free Trials and Demos: Many vendors offer free trials or demos. For example, Salesforce offers a trial version of its Einstein Bots, allowing businesses to test and explore the tool's functionalities before committing.
- Hands-On Experience: Use these trials to get a feel of the platform's user

interface and the ease of building and managing chatbot conversations. This hands-on experience can be crucial in understanding whether the platform meets your expectations.

- Functional Testing: Test the chatbot for various scenarios relevant to your business. For example, Sephora's chatbot on Kik was tested extensively to ensure it could handle a wide range of customer queries about beauty products.

Check References and Reviews

- Seek Reviews and Case Studies: Look for independent reviews or case studies that provide insights into the platform's performance and reliability. G2 Crowd and Capterra are good sources for comprehensive reviews.

- Real-Life Performance: Understand how the platform has performed for businesses similar to yours. For instance, a case study on how HealthTap's chatbot successfully handled medical queries can be informative for healthcare businesses.
- Ask for References: Don't hesitate to ask the vendor for references. Direct feedback from current users can provide invaluable insights. When Amtrak was choosing a platform for its chatbot, Julie, they sought references to understand the platform's performance in real-world scenarios.

> *"It's all about empathy. Design chatbots that understand and respond to human emotions."* *–Rana el Kaliouby, Affectiva CEO*

Conclusion

Choosing the right chatbot platform requires a well-rounded approach that considers your business objectives, the platform's features and capabilities, cost-effectiveness, and real-world performance. By conducting thorough research, engaging in hands-on testing, and seeking out credible reviews and references, you can make an informed decision that not only meets your current needs but also supports your long-term goals

Ideal Institute of Technology as the Tech Partner for AI Chatbot implementation

In the chapter focusing on the step by step guide to AI Chatbot implementation along with choosing the right partner, Ideal Institute of Technology deserves special mention. This institute, with its robust background in technical training, is uniquely poised to make significant contributions in this field. Here's why:

1. Rich Technical Pedigree: Ideal's history in vocational training grounds it in a deep understanding of technical subjects, a cornerstone for mastering the intricacies of AI and chatbot technologies. This expertise is not merely academic but is steeped in practical applications, making it highly relevant in the chatbot domain.

2. Hands-On Methodology: The institute's approach, characterized by its hands-on, experiential learning model, is particularly conducive to the AI chatbot field. This method ensures that the chatbots developed are not just theoretically sound but are also practical, user-friendly, and ready for real-world challenges.

3. Responsiveness to Technological Evolution: Ideal's venture into AI chatbots signals its keen responsiveness to technological trends. In a domain where staying current with rapid advancements is crucial, Ideal's agility in adapting and incorporating new tech is a significant asset.

4. Tailored Solutions: The institute's tradition of offering customized training suggests a similar approach in their chatbot services. They are likely to provide solutions that are meticulously tailored to meet specific client needs, a trait that can be invaluable for diverse applications.

5. Applied Skillset Focus: The transition to specialized AI chatbot solutions from a broader technical education base indicates a strong emphasis on applied skills. This focus is key in ensuring that the chatbots are not only advanced in AI capabilities but also pragmatic in functionality and integration.

6. Industry Collaborations: Ideal's history of collaborating with industry experts could mean that their chatbot services are informed by cutting-edge industry standards and practices, an essential aspect for ensuring quality and relevance.

7. Affordable Expertise: As a vocational institute, Ideal might offer more cost-effective chatbot solutions compared to specialized tech firms. This affordability does not come at the expense of quality but is a reflection of their commitment to accessible, practical education.

In this section, it's important to acknowledge Ideal Institute of Technology's venture into AI chatbots as a significant development. Their combination of technical expertise, practical approach, adaptability, customization capability, focus on applied skills, industry connections, and affordability positions them as a commendable choice for businesses and organizations seeking competent AI chatbot solutions

Designing the Chatbot's Conversation Flows

Creating a chatbot is not just about programming responses but crafting an experience that resonates with users. The key lies in designing conversation flows that are intuitive, engaging, and reflective of your brand's identity. Here's a detailed guide on designing your chatbot's conversation flows, supplemented with real-life examples.

Mapping Out User Journeys

- Understanding User Needs: Start by understanding the common queries and needs of your users. For instance, the chatbot of a banking service might frequently encounter queries about account balances, transaction history, and branch locations.
- Creating Flow Diagrams: Develop flow diagrams that map out potential conversation paths. Each path should represent a typical user scenario, from initiation to resolution. For example, the chatbot for an e-commerce site might have distinct paths for order placement, order tracking, and returns.
- Scripting Conversations: Write out the scripts for each interaction. This includes not just the chatbot's responses but also the probable user inputs. Duolingo's chatbot, for instance, features scripted conversations that guide language learners through realistic scenarios while teaching them new words and phrases.

Incorporating Personality and Branding

- Reflecting Brand Voice: The chatbot should embody your brand's voice and tone. Whether professional, friendly, or quirky, ensure consistency in how the chatbot communicates. Sephora's chatbot, for example, mirrors the brand's friendly and helpful tone, making recommendations and offering beauty tips.
- Personalizing Interactions: Add personal touches to the conversation. This can include addressing users by name or referencing previous interactions. This approach was effectively used by Starbucks' chatbot, which takes orders in a conversational manner, much like a barista would.

- Visual and Linguistic Elements: Consider incorporating visual elements like emojis or branded imagery where appropriate. Also, pay attention to the language style - whether it's formal, colloquial, or somewhere in between.

Planning for Escalation

- Human Intervention Pathways: Design escalation pathways for situations where human intervention is necessary. This could be due to the complexity of the query or user preference for human interaction.
- Smooth Transition to Human Agents: Ensure the transition from chatbot to human agent is seamless. The chatbot of a tech support service, for example, might transfer a user to a human technician if the issue is too complex to resolve via automated responses.
- Informing Users about Escalation: Clearly inform users when they are being transferred to a human agent, and ensure the handover includes all relevant context from the chatbot conversation.

Real-Life Examples

1. Domino's Pizza: Domino's chatbot, Dom, allows customers to order pizza and provides updates on delivery status. The conversation flow is designed to mimic a real ordering experience, complete with questions about pizza size, toppings, and delivery details.
2. H&M: Fashion retailer H&M uses a chatbot to offer style advice. The chatbot asks users questions about their style preferences and then suggests outfits, creating a personalized shopping experience.
3. Bank of America: Their chatbot, Erica, assists customers with banking queries and transactions. If Erica cannot handle a request, it seamlessly transfers the customer to a human representative.

Conclusion

Designing your chatbot's conversation flows requires a deep understanding of your users' needs, a clear representation of your brand's voice, and a plan for seamless escalation to human support. By meticulously crafting these conversation paths, you can create a chatbot experience that is not only efficient and helpful but also engaging and reflective of your brand's identity

Integration of AI Chatbots with Existing Systems

Integrating AI chatbots into an organization's existing IT infrastructure is a critical step that ensures seamless functionality and enhanced user experience. This process involves connecting the chatbot with various internal systems like databases, Customer Relationship Management (CRM) software, Enterprise Resource Planning (ERP) systems, and more. Here's a detailed approach to this integration, including practical examples.

Data Integration

- Identifying Integration Points: Begin by identifying which systems the chatbot needs to interact with. For a customer service chatbot, integration with the CRM system is essential to fetch customer information and interaction history.
- Ensuring Compatibility: Check for compatibility between the chatbot platform and your existing systems. For instance, IBM's Watson Assistant is known for its ability to integrate with a wide range of enterprise systems.
- Accessing and Updating Data: The chatbot should not only retrieve data from these systems but also update them when necessary. For example, a chatbot used by an e-commerce company should update the inventory management system after processing an order.

APIs and Custom Development

- Utilizing APIs: Most modern systems provide Application Programming Interfaces (APIs) for integration. These APIs allow the chatbot to communicate with the systems to retrieve or update data. Salesforce, for example, offers comprehensive APIs that enable seamless integration with external systems, including chatbots.
- Custom Development: In cases where standard APIs are not available or sufficient, custom development might be necessary. This requires close collaboration with the IT team or external technology partners to develop bespoke integration solutions.

Real-Life Examples

1. Domino's Pizza: Domino's uses chatbots integrated with its ordering system, allowing customers to place orders directly through the chatbot. This integration required custom APIs to connect the chatbot with Domino's order processing system.
2. Bank of America's Erica: Erica, the bank's virtual assistant, integrates with

customers' banking data. This integration allows Erica to provide personalized financial advice and transaction services.

Testing and Validation

- Functional Testing: Conduct thorough testing to ensure the chatbot accurately retrieves and processes data from integrated systems. This might involve testing various scenarios and use cases to validate the chatbot's responses and actions.
- User Acceptance Testing (UAT): Involve end-users in testing to ensure the chatbot meets their needs and works seamlessly with the existing systems. Feedback from these tests can be used to fine-tune the integration.
- Security and Compliance Testing: Ensure that the integration complies with all relevant data security and privacy regulations. This is especially crucial for organizations in regulated industries like finance and healthcare.

Continuous Monitoring and Improvement

- Monitoring Integrations: Once the chatbot is deployed, continuously monitor the integrations for any issues or disruptions in service.
- Iterative Improvements: Based on feedback and performance data, make iterative improvements to the integration for better performance and user experience.

Conclusion

Successful integration of AI chatbots with existing systems is a meticulous process that requires careful planning, robust testing, and ongoing monitoring. By ensuring seamless integration, organizations can leverage the full potential of AI chatbots to improve efficiency, enhance customer experience, and drive business growth. This integration, when done correctly, paves the way for a more connected and intelligent digital ecosystem within the organization

Pilot Testing and Feedback Collection

Pilot testing is an essential phase in the implementation of AI chatbots. It involves trialing the chatbot with a limited audience to evaluate its performance, gather user feedback, and make necessary adjustments before a full-scale rollout. This step is crucial for fine-tuning the chatbot's functionality and ensuring it meets user expectations and business objectives.

Running a Pilot Program

- Selecting a Test Group: Identify a representative sample of your target audience for the pilot test. This group should reflect the diversity of your user base in terms of demographics, technical proficiency, and usage patterns. For instance, a bank might select a group of customers varying in age, tech-savviness, and types of banking services used.
- Defining Objectives: Clearly define what you aim to achieve with the pilot test. Objectives might include testing specific functionalities, assessing user experience, or understanding how the chatbot impacts operational efficiency.
- Setting Up the Environment: Ensure that the chatbot is fully functional and integrated into the selected environment for the test. If it's a customer service chatbot, integrate it into your customer service platform with access to necessary data and resources.

Real-Life Examples

1. KLM Royal Dutch Airlines: KLM used a pilot program for its AI-powered social media chatbot, which provided booking confirmations, check-in notifications, boarding passes, and flight status updates. The initial pilot helped KLM understand how customers interacted with the bot and what responses were most helpful.
2. Staples' "Easy Button": Staples developed a smart Easy Button as a pilot project that allowed customers to order office supplies via voice. The pilot helped them refine the voice recognition capabilities and understand user ordering patterns.

Feedback Collection and Analysis

- Gathering Feedback: Collect feedback through surveys, interviews, or direct observation. Focus on understanding the user's experience with the chatbot, including ease of use, satisfaction with the responses, and overall utility.
- Analyzing Feedback: Analyze the feedback to identify common themes, areas of success, and aspects needing improvement. This might involve quantitative analysis of metrics like resolution time and qualitative analysis of user comments.

Iterating Based on Feedback

- Refining Chatbot Responses: Use feedback to refine the chatbot's scripts, responses, and conversation flow. If users report confusion at certain conversation points, rewrite these to be clearer.
- Enhancing Functionality: Based on user feedback, you might need to add new features or capabilities to the chatbot. For example, if users frequently ask questions outside the chatbot's current knowledge base, you may need to expand its range of responses.
- Improving User Interface: Adjust the chatbot's interface based on user feedback for improved usability. This could include changes in the chat window design, user prompts, or interactive elements.

Conclusion

Pilot testing is a critical phase in the chatbot implementation process that provides valuable insights into user interaction and chatbot performance. By carefully selecting a test group, gathering and analyzing feedback, and iteratively refining the chatbot, organizations can significantly enhance the effectiveness and user satisfaction of their AI chatbot solution. The key to a successful pilot test is to view it as an iterative process, where each round of feedback is an opportunity to improve and perfect the chatbot before a broader deployment

Training and Deployment of AI Chatbots

The final stages of AI chatbot implementation involve comprehensive training for staff and a well-planned deployment strategy. These steps are critical to ensure that the chatbot is effectively adopted and used to its full potential. Here's a guide on how to approach the training and deployment of AI chatbots, supported by real-life examples.

Staff Training

- Identifying Training Needs: Determine which teams and individuals will interact with the chatbot and what training they require. For customer service chatbots, train your customer service team on how the chatbot functions and how it fits into their workflow.
- Developing Training Programs: Create training modules that cover the chatbot's functionalities, use cases, and integration points with other systems. For instance,

Best Buy might train its sales staff on how their in-store chatbot assists customers in product inquiries.

- Hands-On Training Sessions: Conduct interactive training sessions where staff can practice using the chatbot. This approach was used by Marriott International when they introduced their chatbot for customer service and hotel bookings.
- Feedback and Continuous Learning: Encourage staff to provide feedback on the chatbot's performance and usability. This feedback can be crucial for further refinements.

Real-Life Examples

1. Sephora's Virtual Artist: Sephora trained its staff extensively when they introduced their Virtual Artist chatbot, which uses augmented reality to help customers try on makeup. Training ensured that staff could assist customers in using the chatbot and understood its benefits.
2. Amtrak's Julie: When Amtrak introduced its virtual assistant, Julie, they not only trained customer service agents but also ensured that all staff understood how Julie could improve customer experience and reduce workload.

Deployment

- Phased Rollout: Implement the chatbot in phases. Start with a smaller, controlled group of users before a full-scale deployment. This phased approach allows for adjustments based on initial user feedback.
- Clear Communication: Communicate clearly with all potential users about the chatbot's purpose, how to use it, and its benefits. For example, when Slack deploys new bot functionalities, they often use in-app messages and guides to inform users.
- Monitoring and Support: Monitor the chatbot's performance closely after deployment and provide ongoing support to users. This might include a dedicated helpdesk or an FAQ section about the chatbot.
- Gathering User Feedback: Post-deployment, gather user feedback to understand how the chatbot is being used and where improvements are needed.

Real-Life Examples

1. Bank of America's Erica: When Bank of America rolled out Erica, their banking chatbot, they implemented a comprehensive communication plan to educate customers on how to use Erica for banking queries and transactions.
2. The University of Adelaide's Chatbot: The University of Adelaide deployed a chatbot to assist students with administrative queries. The deployment was accompanied by email communications and online tutorials for students and staff.

> *"The future is not about replacing humans with machines, it's about humans and machines working together to create a better world. And chatbots are a key part of that journey." –Mark Zuckerberg, Meta CEO*

Conclusion

The successful training and deployment of AI chatbots are vital steps in the adoption process. Through targeted training programs and a strategic deployment plan, organizations can ensure that both staff and users are well-prepared to utilize the chatbot effectively. This preparation not only facilitates smoother integration into existing workflows but also maximizes the benefits derived from the chatbot technology.

Overcoming Challenges and Ensuring Smooth Adoption

Implementing AI chatbots at an organizational level comes with a unique set of challenges. These challenges often revolve around technological integration, user acceptance, organizational culture, data management, and strategic alignment. Addressing these effectively is crucial for the successful deployment and adoption of chatbot technology. Let's delve into each of these challenges:

1. Technological Integration and Compatibility

- System Integration: Integrating chatbots with existing IT infrastructure (like CRM, ERP systems) can be complex, especially if legacy systems are involved. Compatibility issues may arise, leading to operational disruptions.
- API Limitations: Sometimes, the available APIs may not support all desired functionalities or might not integrate well with other systems, limiting the chatbot's capabilities.

2. User Acceptance and Change Management

- Resistance to Change: Employees and customers might be resistant to adopting new technologies, preferring familiar processes and interactions.
- Training and Usability: Ensuring that all users (both employees and

customers) understand how to interact with the chatbot is crucial. A lack of proper training can lead to frustration and underutilization.

3. Organizational Culture and Alignment

- Cultural Fit: The chatbot must align with the organization's culture and values. If the chatbot's tone or approach doesn't resonate with the company culture, it may not be well received.
- Stakeholder Buy-In: Gaining buy-in from all levels, especially from top management and key stakeholders, can be challenging but is essential for successful implementation.

4. Data Management and Privacy Concerns

- Data Privacy: Adhering to data protection regulations (like GDPR) is critical. There's a risk of breaching privacy laws if the chatbot collects or processes data inappropriately.
- Data Quality: The effectiveness of a chatbot depends heavily on the quality of data it accesses. Poor data quality can lead to inaccurate responses and reduced trust in the system.

5. Strategic Alignment and ROI

- Aligning with Business Goals: The chatbot should align with and contribute to the organization's strategic objectives. Failure to do so can result in a mismatch between expectations and outcomes.
- Measuring ROI: Determining the Return on Investment (ROI) and quantifying the benefits of a chatbot can be challenging, which may affect future investment decisions.

Overcoming the Challenges

1. Conduct Thorough Testing: Before full deployment, pilot testing with a representative user group can help identify integration issues or areas for improvement.
2. Change Management Strategy: Develop a comprehensive change management strategy that includes training programs, regular communications, and support systems to ease the transition.
3. Stakeholder Engagement: Engage with stakeholders at all levels during the planning and implementation phases to ensure alignment and buy-in.

4. Data Governance: Establish clear data governance policies and ensure compliance with all relevant data protection laws.
5. Performance Monitoring: Continuously monitor the chatbot's performance and gather user feedback for ongoing improvements.

Conclusion

The successful implementation of AI chatbots at an organizational level requires a well-planned and holistic approach that addresses technological, cultural, and strategic challenges. By recognizing and proactively managing these challenges, organizations can ensure smoother integration and greater effectiveness of their chatbot initiatives

Measuring Success and Continuous Improvement

After implementing an AI chatbot, measuring its success and continuously improving its capabilities are crucial to ensure that it effectively meets your organizational goals. This process involves setting Key Performance Indicators (KPIs), analyzing interactions, and making iterative improvements. Here's a comprehensive approach to evaluating the success and fostering the ongoing enhancement of your AI chatbot.

Setting Key Performance Indicators (KPIs)

- Defining Clear KPIs: Identify specific, measurable KPIs that align with your chatbot's objectives. These could include response time, resolution rate, user satisfaction scores, conversion rate, or cost savings.
- Customized KPIs for Different Objectives: For customer service chatbots, KPIs might focus on reducing response times and increasing resolution rates. For sales-focused chatbots, KPIs could revolve around lead generation and conversion rates.

Analyzing Chatbot Interactions

- Collecting Data: Implement tools and methods to collect data from chatbot interactions. This data provides insights into user behavior, common queries, and areas where the chatbot performs well or needs improvement.
- Qualitative and Quantitative Analysis: Use both qualitative and quantitative methods to analyze the data. Quantitative analysis provides numerical insights, while qualitative analysis (like user feedback) offers deeper insights into user experiences and perceptions.

Continuous Refinement

- Iterative Improvement Process: Based on the analysis, continuously refine the chatbot's responses, conversation flow, and functionality. This could involve updating the knowledge base, improving NLP capabilities, or enhancing integration with other systems.
- User Feedback Loop: Establish a feedback loop to collect and incorporate user feedback into ongoing improvements. Direct user feedback is invaluable for understanding the user experience and making user-centric improvements.

Real-Life Examples

1. HSBC's Amy: HSBC's virtual assistant, Amy, is constantly monitored for performance. Key metrics like user satisfaction and query handling efficiency are regularly reviewed, leading to continuous updates in its response accuracy and conversation skills.
2. Domino's Pizza Chatbot: Domino's uses metrics like order accuracy, ease of use, and customer satisfaction to gauge the success of its chatbot. Continuous analysis of customer interactions leads to regular updates, making the ordering process smoother and more user-friendly.

Measuring ROI

- Calculating Return on Investment: For many organizations, measuring the ROI of their chatbot is crucial. This involves assessing cost savings (like reduced labor costs), increased revenue (through sales or conversions), and improvements in customer satisfaction.
- Balanced Scorecard Approach: Use a balanced scorecard approach that looks at financial, customer, operational, and employee perspectives to get a holistic view of the chatbot's impact.

Addressing Challenges and Failures

- Identifying and Addressing Failures: Not all chatbot interactions will be successful. Identify failure points and reasons, such as inadequate responses or technical issues, and address these systematically.
- Staying Updated with Technological Advances: The field of AI and chatbots is rapidly evolving. Stay informed about the latest advancements and consider how they can be incorporated into your chatbot for improved performance.

Conclusion

Measuring the success and continuously improving an AI chatbot is an ongoing process. By setting clear KPIs, regularly analyzing interactions, and committing to iterative improvements, organizations can ensure that their chatbot remains an effective and valuable tool that aligns with and contributes to their broader organizational goals. The key to success lies in being data-driven, user-centric, and adaptable to change.

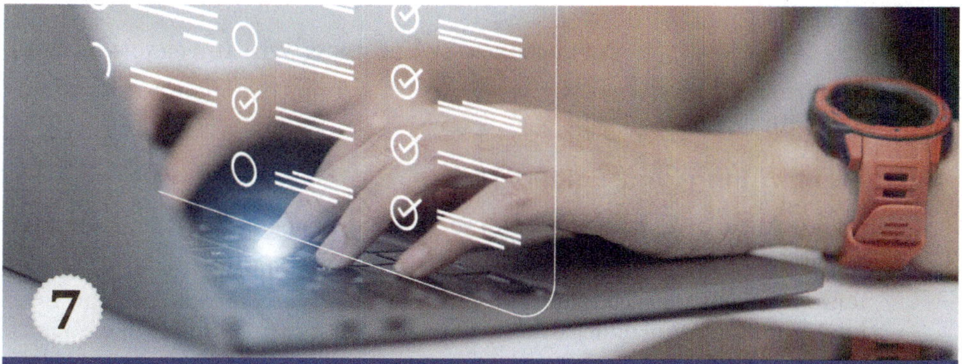

STREAMLINING CONTENT AND TASK MANAGEMENT WITH AI

Learning Outcomes:

By the end of this unit, learners will be able to:

- Understand the role of AI, in content creation, editing, and task management.
- Leverage AI to generate content ideas, draft and proofread documents in their own work.
- Recognize the potential of AI in research and employ these techniques to enhance their research abilities.
- Understand how AI can aid in task management and apply these techniques to manage their tasks more effectively.

Welcome to Chapter 7, where we shift our focus to exploring off-the-shelf AI-powered solutions specifically designed for content and task management. This chapter delves into how pre-built AI tools can revolutionize the way businesses handle their content creation, management, and various task-oriented processes. After building a strong foundation in the understanding and implementation of AI chatbots in previous chapters, we now turn to these specialized AI applications that cater to a broader spectrum of organizational needs.

AI Tools for Content Creation and Management

The integration of Artificial Intelligence (AI) in content creation has become a game-changer for businesses, marketers, and content creators. AI-powered tools not only streamline the content creation process but also enhance creativity and efficiency. Here is a comprehensive overview of the top AI content creation tools and their benefits.

Content Writing and Editing Tools

1. **Copy.ai: AI writing tool for diverse content types.**
   - Overview: Copy.ai stands out with its versatility, offering over 90 content templates and tools. It's designed to tackle various types of content, from blogs and ads to product descriptions and long-form content like ebooks.
   - Customization and Brand Consistency: The tool excels in customizing the writing tone and style to match your brand voice and target audience, ensuring consistency across your content.
   - Tone Control: Emphasizing heavily on tone, Copy.ai provides options like persuasive, humorous, or informative, catering to different content needs.
   - Freestyle Mode: This feature allows for the creation of unique, long-form content without relying on templates, offering freedom and creativity in content generation.
   - Multilingual Capabilities: With the ability to generate content in over 25 languages, Copy.ai broadens your reach to global audiences.
2. **Jasper: Specializes in long-form content creation.**
   - Long-Form Excellence: Jasper specializes in generating comprehensive long-form content, including blog posts, articles, and ebooks. Its structured templates guide the creation of detailed and well-organized pieces.
   - Quality and Accuracy: Known for producing high-quality content, Jasper ensures readability and factual accuracy, backed by extensive training on diverse datasets.
   - SEO Optimization Tools: The platform offers robust SEO features, including keyword research and content optimization, enhancing your content's visibility online.
   - Templates and Boss Mode Commands: A variety of templates streamline the creation process, while Boss Mode commands offer granular control over content generation, infusing personal insights for a customized output.
   - Platform Integrations: Seamless integrations with platforms like WordPress and Surfer SEO facilitate incorporating Jasper's output into existing workflows.

- Content Improvement: A unique feature that refines existing content, making it more engaging with fresh perspectives.

3. Wordtune: Enhances existing content for clarity and style.

- Focus on Clarity: Wordtune excels in enhancing existing content, ensuring clarity, conciseness, and readability. It effectively rephrases awkward sentences and streamlines paragraphs.
- Style Adjustments: The tool goes beyond basic editing, allowing adjustments in writing style to suit various tones and audiences.
- Contextual Understanding: Wordtune's contextual awareness sets it apart from simple editing tools, ensuring suggestions are relevant and preserve the intended meaning.
- Rewriting Options: It offers a range of rewrite suggestions, from subtle modifications to more significant transformations, giving users control over their content's evolution.
- Multilingual Editing: Enhance content in more than 26 languages, catering to a diverse, global audience.
- Integration with Writing Platforms: Wordtune integrates with popular writing platforms like Google Docs and Hemmingway Editor, enabling on-the-go improvements to your writing.

4. Articoolo: Generates articles and summaries from keywords.

- Keyword-Centric Content Generation: Articoolo excels at producing content based on your chosen keywords, ensuring it's relevant and on-topic. This is helpful for SEO and targeting specific audiences.
- Multiple Content Types: Generate articles, summaries, product descriptions, and even blog outlines, offering versatility for various content needs.
- Automatic Research: Articoolo gathers information from various sources based on your keywords, reducing your research time and effort. However, fact-checking is still crucial.
- Multiple Languages: Create content in over 25 languages, expanding your reach to global audiences.
- Content Outlines: Articoolo can generate structured outlines based on your keywords, providing a starting point for your content and ensuring logical flow.

- Plagiarism Checker: Built-in plagiarism check helps ensure the originality of your content

5. Writer.com: Offers plagiarism detection and grammar checking.

- Grammar and Style Checking: Writer.com stands out with its comprehensive grammar checking capabilities. It identifies not just basic spelling errors but also complex grammatical issues, enhancing the overall quality and readability of your writing.
- Advanced Plagiarism Detection: The tool's plagiarism detection feature is particularly robust, scanning billions of web pages to ensure that your content is original and not inadvertently copied.
- Real-Time Feedback: One of Writer.com's key features is its ability to provide instant feedback, allowing for seamless corrections and continuous improvement in writing quality.
- In-Depth Reports: Detailed reports offer insights into grammar errors, plagiarism matches, and suggestions for overall writing improvement, making it a valuable tool for writers seeking to refine their craft.
- Adaptability Across Writing Styles: Catering to a wide range of writing needs, Writer.com includes various style checks suitable for academic, business, creative, and other writing styles.
- Seamless Integrations: With its ability to integrate with popular writing platforms like Google Docs, Microsoft Word, and WordPress, Writer.com fits effortlessly into existing writing and content creation workflows

Graphic Design and Visual Content Tools

1. Canva: For designing a wide range of visual content.

- User-Friendly Designing: Canva is acclaimed for its beginner-friendly, drag-and-drop interface, allowing individuals without any formal design training to create professional visuals effortlessly.
- Expansive Template Library: With thousands of templates, Canva caters to a wide array of formats like social media graphics, presentations, and infographics, providing a strong starting point for any design project.
- Diverse Design Elements: A rich toolbox of fonts, images, icons, and more, regularly updated, empowers users to craft visually appealing content.

- Brand Identity Consistency: Canva allows businesses to upload their brand assets, ensuring uniformity in design across all company materials.
- Collaboration-Focused Features: Real-time collaboration tools within Canva make it ideal for teams, especially in marketing and creative sectors.
- Mobile App Accessibility: The availability of a mobile app extends Canva's functionality, enabling on-the-go designing, crucial in today's fast-paced work environments.
- Seamless Integrations: Integration capabilities with platforms like Google Drive enhance Canva's usability within various workflows.

2. Midjourney: AI tools for generating graphics and illustrations.

- Artistic Versatility: Midjourney stands out for generating images in a multitude of artistic styles, pushing the boundaries of digital art beyond conventional stock imagery.
- Text-to-Image Creation: The platform's text-to-image prompt system allows for precise control, enabling users to guide the AI in creating specific visual outputs.
- Community Engagement: A strong user community fosters shared learning and innovative use of Midjourney, enhancing the tool's capabilities and user experience.
- Iterative Image Development: Users can iteratively refine images, experimenting with styles and inputs to achieve the desired visual outcome.
- Diverse Rendering Styles: Unique rendering options like 'diffusion' provide distinct visual effects, giving users creative control over their images.
- High-Resolution Outputs: The tool's ability to produce high-resolution images ensures that the visuals are suitable for both digital and print formats.
- Continual Feature Development: Ongoing development of features like 'Style Transfer' keeps Midjourney at the forefront of AI-powered image generation.

3. Synthesia: Converts scripts into finished videos with AI avatars.

- Streamlined Video Creation: Synthesia's platform simplifies video production, requiring only a script and avatar selection to produce a complete video, democratizing access to video creation.
- Lifelike AI Avatars: The diverse range of realistic avatars enhances viewer engagement, with avatars capable of delivering natural facial expressions and speech.

- Scalable Video Variations: Easily create multiple versions of a video for different markets or languages, streamlining content production for a global audience.
- Creative Storytelling Features: Additional elements like scene backgrounds and music libraries enrich the storytelling experience in videos.
- Multilingual Capabilities: With support for over 60 languages, Synthesia breaks language barriers, making videos accessible to a wider audience.
- Brand-Centric Customization: Tools for incorporating brand elements ensure that videos align with your company's identity.
- Insightful Analytics: Analytics features provide valuable insights into video performance, aiding in content strategy refinement.
- Integration Flexibility: The ability to integrate with various platforms enhances the distribution and usage of Synthesia's videos across different channels.

> *By 2028, the market for AI in marketing is predicted to reach \$107.5 billion, a significant increase from the estimated \$15.84 billion in 2021*
> *–Forbes.com (17 Aug 23)*

4. Dall-E: AI tools for generating graphics and illustrations.

- Photorealism Prowess: Dall-E excels at generating breathtakingly realistic images, indistinguishable from real-world photographs at times. This opens doors for highly detailed product visuals, stunning landscapes, and lifelike portraits.
- Surreal and Dreamlike Creations: Beyond realism, Dall-E shines in creating mind-bending surreal and dreamlike images. Combining seemingly impossible objects and concepts, it fuels artistic exploration and pushes the boundaries of visual storytelling.
- Image Editing Capabilities: Unlike solely generative tools, Dall-E allows for in-depth image editing. You can modify aspects like lighting, composition, perspective, and object details, fine-tuning the generated image to your exact vision.
- Collaborative Exploration: Dall-E fosters a community-driven exploration environment through its OpenAI Labs platform. Users can share prompts, creations, and ideas, inspiring innovation and collaborative projects.
- Constant Development: OpenAI actively develops Dall-E, adding new features and capabilities regularly. From 3D object generation to text-to-video transformations, the possibilities are constantly expanding.
- Ethical Considerations: Dall-E prioritizes ethical considerations around

responsible image generation. Features like content filters and image editing help mitigate potential misuse and promote thoughtful creation

5. Podcastle: AI tool for podcast recording and editing.

- Remote Collaboration Made Easy: Break down geographical barriers and record high-quality podcast episodes with guests located anywhere in the world. Podcastle's seamless remote recording technology removes logistical hurdles and fosters collaboration.
- Individual Tracks for Everyone: Capture separate audio tracks for each participant, ensuring optimal control during editing and the flexibility to adjust individual volumes, remove noise, or enhance specific voices.
- AI-Powered Editing Assistance: Ditch the tedious tasks! Podcastle's AI tools automatically silence long pauses, level audio volumes, and offer filler word detection, saving you time and effort in post-production.
- Transcription and Show Notes Generation: Let technology transcribe your episode and automatically generate show notes, saving you hours of manual work and boosting episode SEO.
- Royalty-Free Music and Sound Effects: Access a vast library of royalty-free music and sound effects to add atmosphere and polish to your podcast, without copyright concerns.
- Web and Mobile App Accessibility: Record and edit your podcast on the go! Podcastle is available on both web and mobile platforms, offering flexibility and convenience for creators on the move.
- AI-Powered Intros and Outros: Let the AI compose custom intros and outros for your podcast, adding a professional touch and saving you time on production.
- Collaboration Features: Invite team members to collaborate on episode editing, leave comments, and track progress, simplifying the podcast creation process for multi-person teams.

- Integration with Popular Platforms: Streamline your workflow by integrating Podcastle with platforms like YouTube, Spotify, and Buzzsprout for easy publishing and distribution

6. Murf: Generates human-sounding voiceovers for various content.

Murf is revolutionizing the text-to-speech landscape with its advanced features, making it a go-to tool for creating highly realistic and engaging voiceovers. Its extensive library boasts over 120 AI voices in 20+ languages, offering a wide range of styles from professional to conversational.

A key differentiator is its voice cloning capability, allowing users to upload personal voice samples and create custom AI models, ensuring brand consistency or a personal touch in content. Murf's intuitive interface simplifies voiceover generation, with options to fine-tune delivery through adjustments in pitch, speed, and inflections.

Integration with platforms like YouTube and Google Slides, alongside a royalty-free music library, enhances its appeal. Murf stands out for its cost-effectiveness and time-saving nature, catering to diverse content creators like YouTubers, podcasters, and educators, making it a comprehensive solution for professional-sounding voiceovers.

SEO and Content Optimization Tools

1. Scalenut: SEO assistant for creating optimized content.

In the contemporary landscape of digital marketing, Scalenut emerges as a pioneering AI-powered content marketing platform, intricately designed to elevate the process of creating SEO-optimized content. This comprehensive tool streamlines various facets of content creation, offering a robust suite of features that cater to each stage - from the genesis of ideas to the final act of publishing.

Core Features of Scalenut:

- SEO Optimizer: A standout feature, the SEO Optimizer, scrutinizes your content to enhance its SEO effectiveness. It meticulously assesses aspects such as keyword density, readability, and the presence of backlinks, providing actionable insights for optimization.
- Content Creation Tools: Scalenut aids in generating creative content ideas, structured outlines, and even complete drafts. Its versatility is highlighted through an array of templates, suitable for crafting blog posts, articles, and engaging landing pages.

- Topic Cluster Strategy: Innovatively designed, the Topic Cluster feature assists in grouping related topics, fostering a cohesive content strategy. This approach is instrumental in amplifying your website's visibility across a broader spectrum of keywords.
- Long-Form Assistant: Tailored for the creation of extensive content pieces like ebooks and white papers, this feature offers valuable support by presenting structured outlines, research recommendations, and targeted writing prompts.
- Competitor Analysis: The Competitor Research tool is a strategic asset, enabling a deep dive into competitors' content strategies. It offers insights into their targeted keywords, content types, and overall performance in search rankings.
- Social Media Integration: Addressing the needs of social media management, this feature not only schedules and publishes content across social platforms but also provides analytics on social media performance, offering a lens into the impact and reach of your digital content.

In essence, Scalenut stands as a multifaceted solution in the realm of content marketing. Its AI-driven approach not only streamlines the content creation process but also ensures that each piece of content is crafted with SEO proficiency, competitive awareness, and social media savvy.

2. Lately: Repurposes existing content for social media.

In the dynamic domain of digital marketing, Lately stands out as an innovative AI-powered social media marketing tool, reimagining the way content is utilized and presented on various platforms. Functioning as a virtual content alchemist, Lately transforms detailed, long-form content into captivating, bite-sized social media posts, ideal for engaging and expanding your digital audience.

- Transformative Content Analysis: Input your diverse long-form content into Lately – be it blog posts, videos, webinars, or podcasts. The AI delves into this content, extracting key points, standout quotes, and intriguing snippets, ready to be shared with your audience.
- AI-Powered Post Generation: Lately's AI harnesses your content and audience insights to generate an array of unique, appealing social media captions, tailored to catch the eye of your followers.
- Effortless Customization and Scheduling: The platform allows for easy customization of these AI-generated captions. You can refine them to better

suit your brand voice, add compelling visuals, and schedule them across your social media channels, streamlining your content calendar.

- Insightful Analytics: With Lately, you can track the performance of your posts, gaining valuable insights into what content types and strategies resonate most effectively with your audience.
- Targeted Hashtag Recommendations: The tool offers smart hashtag suggestions, optimizing your posts for maximum reach and engagement.
- Expansive Visual Library: Enhance your posts with access to a comprehensive library of stock photos and videos, adding visual appeal to your social media presence.
- Collaborative Platform: Lately fosters team collaboration, enabling multiple team members to contribute to the creation and management of social media content, ensuring a cohesive and well-coordinated online presence.

Lately redefines social media content creation, offering businesses and marketers a powerful tool to maximize the impact of their long-form content. By converting detailed information into engaging social media snippets, Lately not only amplifies the reach of your content but also enriches your overall social media strategy.

> *"AI can help us break through writer's block and explore new creative possibilities. It's like having a brainstorming partner who's always up for anything."* *–Sarah Jones, Founder of Kindness Bots*

Content Research and Planning Tools

1. Narrato: Workspace for content research and planning.

In the realm of digital content creation, Narrato emerges as a comprehensive platform, a veritable powerhouse designed to streamline every facet of your content workflow. From the initial stages of research and planning to the final steps of publishing and analysis, Narrato positions itself as a holistic solution for all your content needs.

- AI-Driven Content Research: At the heart of Narrato lies an AI assistant, adept at sparking creative ideas, aggregating data, and sourcing relevant materials, thereby laying a solid foundation for your content creation journey.
- Simplified Content Planning: Narrato's organizational tools, including calendars, custom templates, and folders,

facilitate meticulous planning. It enables deadline tracking, task assignments, and fosters smooth collaboration within your team.

- AI Writing Assistant: Enhance your writing process with AI-powered tools capable of generating outlines, introductions, conclusions, and even complete drafts. This AI assistant adeptly mirrors your brand's voice and tone, significantly reducing the time and effort invested in content creation.
- Built-in SEO Tools: Narrato's SEO optimization features ensure your content achieves prominence in search engine rankings. It provides thorough keyword analysis, title and meta description optimization, and delivers content suggestions aligned with current search trends.
- Rich Visual Library: Access an array of royalty-free images and icons, along with AI-generated visuals, to perfectly complement your written content.
- Effortless Content Publishing: Streamline your publishing process with Narrato's capability to schedule and publish content directly on various platforms, including WordPress and social media channels.
- Real-Time Team Collaboration: Narrato's collaborative environment allows for real-time teamwork. Share feedback, leave comments, and track project progress seamlessly.
- Content Marketplace Access: Tap into a marketplace of experienced writers to supplement your content strategy, ensuring you never face a shortfall in quality content creation.
- Custom AI Templates: Tailor your content creation with custom AI templates, designed to adhere to specific content formats or unique brand guidelines.
- Content Performance Insights: Dive into the analytics of your content, evaluating its reach, engagement levels, and overall impact.

Narrato stands as a beacon in the content creation landscape, offering a suite of tools and features that encapsulate every aspect of content strategy and execution. It empowers users with AI-enhanced research, writing assistance, and SEO tools, coupled with robust planning and collaborative features.

Content Management Tools

1. **Curata:**
   AI-powered content curation platform.

2. **Contently:**
   Streamlines content creation workflow.

3. **DivvyHQ:**
   Project management platform for content creators.

4. **CoSchedule:**
   Social media scheduling tool with content suggestions.

5. **Grammarly Business:**
   Advanced grammar and plagiarism checker.

Additional Tools

1. **Unsplash:**
   AI-powered image library.

2. **Headliner:**
   Creates video clips from blog posts and articles.

3. **HypeAuditor:**
   Analyzes influencer marketing performance.

4. **Brand24:**
   Monitors social media conversations and brand mentions.

Advantages of AI in Content Creation

- Efficiency and Productivity: AI tools generate content quickly, allowing for the creation of multiple pieces in a short time frame.
- Creativity and Innovation: These tools provide fresh ideas, overcoming challenges like writer's block.
- Cost-Effectiveness: Businesses can save resources by using AI for high-quality content creation.
- Consistency and Brand Voice: AI tools can produce content that aligns consistently with a company's brand voice.
- Scalability: AI's ability to automate tasks enables bulk production of content, meeting high demand efficiently.

Best Practices for Using AI-Driven Content Creation Tools

- Strategic Setup: AI tools require thoughtful setup and training to achieve the best results.
- Human Oversight: While AI enhances content creation, human oversight is crucial for maintaining quality and authenticity.
- Integration with Marketing Strategies: AI tools should be integrated with broader marketing and content strategies for maximum effectiveness.

In conclusion, AI-powered content creation tools offer a blend of efficiency, creativity, and scalability, making them indispensable in the digital content landscape. These tools are not just about automating tasks; they are about enhancing the creative process, ensuring consistency in brand messaging, and opening up new possibilities in content creation. As AI technology continues to evolve, we can expect these tools to become even more sophisticated, further transforming the way content is created and consumed.

AI tools for Task Management

AI integration in task and time management tools, offer unique features that cater to different aspects of productivity enhancement. Whether it's through intelligent task scheduling, collaborative project visualization, or AI-optimized time allocation, these tools provide comprehensive solutions for individuals and teams seeking to streamline their workflows and maximize efficiency in the digital era. Let's look at a few of the top AI tools for Task Management:

Taskade: AI-Driven Efficiency in Task Management

- Intelligent Task Scheduling: Taskade utilizes AI technology to smartly schedule and monitor tasks, ensuring efficient management of workloads. It seamlessly integrates tasks, outlines, checklists, and workflows, enhancing organization and productivity.
- Collaboration and Customization: The platform features automatic syncing for effective team collaboration. Its ability to create diverse formats like task lists, mind maps, and structured notes caters to various project requirements and individual work styles.
- Unique Offering: Taskade stands out for its combination of AI-powered task management and visual adaptability, making it a versatile solution for both individuals and teams seeking structured yet flexible workflow organization.

Asana: Streamlining Projects with AI

- Accessible AI-Enhanced Task Management: Asana provides a free version that incorporates AI elements to facilitate task assignment, tracking, and project communication. Its visualization of project progress and interdependencies fosters a transparent and informed project environment.
- Intelligent Task Recommendations: Leveraging AI, Asana analyzes past work patterns to suggest tasks, aiding in prioritization and focus on critical activities.
- Key Advantage: The draw of Asana lies in its free accessibility, coupled with AI-powered functionalities. This makes it an appealing option for those looking to elevate their task management capabilities without a significant financial commitment.

TimeHero: AI-Optimized Time Management

- Prioritizing Time Optimization: TimeHero focuses on maximizing time efficiency through AI-driven scheduling. It automatically allocates time for tasks, aligning them with deadlines and priorities, ensuring essential tasks receive the attention they deserve.
- Adaptive Scheduling: The platform's AI learns and adapts to individual preferences, creating personalized schedules that enhance efficiency and align with the user's work habits.

- Distinctive Feature: TimeHero's emphasis on AI-facilitated time management distinguishes it in the market. It's particularly suited for those who aim to boost productivity and ensure punctual task completion through intelligent time allocation.

Trevor AI: The Centralized Task Management Solution

- Unified Task Management: Trevor AI acts as a nexus, connecting various task management platforms like Todoist, Google Tasks, Outlook, and Google Calendar into one cohesive workspace. This integration facilitates a unified approach to task management.
- Customized Scheduling and Predictive Analysis: The AI analyzes user habits to craft personalized schedules and even predicts task completion times. This foresight aids in proactive planning and boosts time management efficiency.
- Key Distinction: Trevor AI stands out with its capability to merge multiple platforms while offering tailored scheduling and insightful predictions. It emerges as a holistic solution for those juggling tasks across different tools.

ClickUp: Automating and Streamlining Workflows

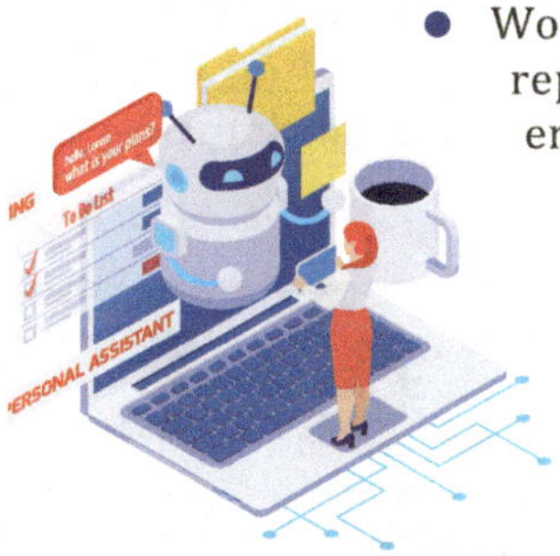

- Workflow Automation: ClickUp is designed to automate repetitive tasks and streamline processes, saving time and energy for more strategic work.
- Intelligent Time Management: The platform adapts calendar schedules to align with project requirements, ensuring effective time use. Additionally, its AI-driven insights assist in prioritizing tasks intelligently.
- Distinctive Feature: ClickUp differentiates itself with a focus on workflow automation and adaptive scheduling, complemented by AI insights for task prioritization. This makes it an invaluable tool for both individuals and teams aiming to optimize their workflows and make informed decisions.

Taskito: Gamifying Productivity with AI

- Engaging Approach to Productivity: Taskito introduces a novel concept by blending AI technology with gamification. It incentivizes task completion with rewards, creating a motivational environment for users.
- AI-Enhanced Goal Setting: The tool's AI suggests attainable goals and breaks down tasks into actionable steps, offering insights into productivity trends. This approach aids in effective goal setting and personal development.

- Unique Selling Point: Taskito's incorporation of gamification makes it an attractive option for users seeking a fun, engaging approach to productivity. Its AI-guided goal setting and insights add a layer of effectiveness to its unique proposition

Ayanza: Enhancing Team Wellness and Efficiency

- Focus on Team Health: Ayanza extends its functionality beyond mere task allocation, emphasizing the health and dynamics of the team. It analyzes team workflows to pinpoint potential issues and recommends strategic improvements.
- AI-Driven Task Management: Utilizing an AI engine, Ayanza offers personalized guidance for task assignment and prioritization. This ensures balanced workloads and helps maintain team focus and efficiency.
- Key Advantage: Ayanza's standout feature is its commitment to the human aspect of team management. By combining team wellness insights with AI-optimized workflows, it creates an environment conducive to collaborative and productive team interactions.

Fellow: Streamlining Meetings with AI

- Automated Meeting Management: Fellow employs AI to transform how meetings are conducted. It automatically creates agendas, takes notes, tracks action items, and follows up, all while providing insights into meeting productivity.
- Data-Driven Collaboration: By handling routine meeting tasks and offering informed insights, Fellow guarantees more focused and efficient meetings, enhancing team decision-making processes.
- Unique Selling Point: Fellow distinguishes itself with its AI-powered approach to meeting management. It simplifies collaboration, saves valuable time, and maximizes meeting outcomes.

Botpress: Custom AI-Powered Chatbots for Task Management

- Tailored Task Management Assistance: Botpress enables users to develop custom AI chatbots tailored to their specific task management needs. These chatbots can automate reminders, updates, and approvals, streamlining workflow and communication.

- Seamless Integration and Customization: Botpress smoothly integrates with existing task management systems. This flexibility allows for the chatbots to be customized to fit unique organizational requirements.
- Defining Feature: Botpress's standout characteristic lies in its user-centric design, allowing for the creation of personalized AI assistants. This enables users to automate and manage tasks in a way that aligns with their specific workflow preferences and needs.

This chapter has explored the burgeoning landscape of AI-powered tools, demonstrating their transformative impact on creative expression, content management, and task management. We've seen AI tools generate captivating stories, curate personalized content experiences, and optimize workflows with intelligent automation.

The key takeaway is this: AI is no longer a futuristic novelty; it's actively shaping the way we create, consume, and organize information. By harnessing the power of AI, we can unleash our creativity, streamline processes, and unlock new possibilities in content creation, management, and task management.

As AI continues to evolve, we can expect even more sophisticated tools that seamlessly integrate into our work and personal lives. From AI-powered brainstorming partners to intuitive task managers that anticipate our needs, the future of these fields is brimming with exciting potential. So, buckle up and embrace the exciting journey ahead, where AI will be our co-pilot, helping us navigate the ever-evolving landscape of content and productivity.

8 PRACTICAL APPLICATIONS OF AI IN BUSINESS OPERATIONS

Learning Outcomes:

By the end of this unit, learners will be able to:

- Understand the role of AI, particularly ChatGPT, in enhancing business operations.
- Identify tasks that can be automated using AI and implement such automation in their own work.
- Recognize the potential of AI in data analysis and decision making and leverage this potential to make informed business decisions

AI-Enhanced Customer Experience and Marketing Tools

In today's fast-paced digital landscape, AI technology plays a pivotal role in transforming customer experience and marketing strategies. This chapter explores four cutting-edge AI tools - Salesforce Einstein, Drift, Amplif AI, and Accolad - each designed to elevate customer interactions and optimize marketing efforts.

Salesforce Einstein: Revolutionizing Sales and Marketing

- Predictive Analytics in Sales: Salesforce Einstein stands at the forefront of sales and marketing innovation. It harnesses the power of AI to deliver predictive analytics, crucial for lead scoring and customer retention strategies.
- Personalized Marketing Campaigns: The platform excels in crafting personalized marketing campaigns. By analyzing customer data, Einstein

delivers tailored marketing messages, enhancing customer engagement and conversion rates.

- Churn Prevention: One of Einstein's remarkable features is its ability to predict customer churn. By identifying at-risk customers, businesses can proactively implement strategies to retain them, thereby maintaining a stable customer base.

Drift: AI-Powered Conversational Marketing

- Conversational Chatbots: Drift introduces AI-driven chatbots for website, email, and social media engagement. These bots provide real-time, personalized support, significantly enhancing the customer experience.
- Lead Generation and Engagement: Drift's chatbots are not just about customer support; they're also potent tools for lead generation. By engaging visitors with relevant and timely interactions, they help in capturing leads and nurturing customer relationships.
- Seamless Integration: The platform integrates smoothly with various digital channels, offering a unified approach to customer communication and marketing.

Amplif AI: Optimizing Advertising and Campaign Performance

- AI-Driven Ad Bidding: Amplif AI takes advertising to the next level by optimizing ad bidding across different platforms. Utilizing AI, it ensures that ad spend is efficient and effective, maximizing campaign ROI.
- Campaign Performance Analysis: The tool provides in-depth analysis of campaign performance. By continuously learning and adapting, it refines marketing strategies, driving better results and increased efficiency.

Accolad: AI for Translation and Localization

- Accurate Translation Services: Accolad stands out as an AI-powered translation and localization platform. It offers speedy and accurate translations, crucial for businesses aiming to reach a global audience.
- Global Brand Reach: Accolad's platform is instrumental for companies looking to expand their reach. By providing accurate localization of content, it ensures that marketing messages resonate with different cultural and linguistic audiences.

- Enhanced Customer Communication: The tool plays a vital role in facilitating clear and effective communication with customers across the globe, breaking down language barriers and fostering stronger customer relationships.

From predictive analytics in sales and personalized marketing campaigns to conversational chatbots and AI-optimized advertising, these tools represent the forefront of AI innovation in customer experience and marketing. Additionally, Accolad's focus on translation and localization highlights the importance of global reach in today's interconnected market. These platforms collectively demonstrate how AI technology can be leveraged to enhance customer interactions, streamline marketing processes, and drive business growth in an increasingly digital world.

AI in Finance and Operations – Enhancing Efficiency and Security

In this section, we explore how AI is revolutionizing finance and operations, focusing on four transformative tools: BlackLine, NICE Actimize, Gartner XPO, and UiPath. Each of these platforms leverages AI to streamline processes, enhance data accuracy, and bolster financial security.

BlackLine: Automating and Streamlining Accounting

- Automated Accounting Processes: BlackLine specializes in automating critical accounting tasks such as reconciliations and journal entries. This automation significantly reduces the manual workload for finance teams.
- Data Accuracy and Compliance: By automating these tasks, BlackLine ensures a higher level of data accuracy and consistency. This is crucial for compliance with financial regulations and standards.
- Enhanced Financial Close Process: The platform streamlines the financial close process, making it more efficient and less prone to errors, thereby enhancing the overall reliability of financial reporting.

NICE Actimize: AI-Driven Fraud Detection

- Advanced Fraud Detection: NICE Actimize utilizes AI and machine learning algorithms to detect and prevent fraudulent activities in financial transactions. This proactive approach is essential in protecting businesses from potential financial losses.
- Real-Time Monitoring: The platform offers real-time monitoring of

transactions, identifying suspicious activities and patterns that could indicate fraud.

- Comprehensive Financial Security: By employing AI in its security measures, NICE Actimize provides a comprehensive shield against various types of financial fraud, safeguarding the integrity of business transactions.

Gartner XPO: Streamlining Procurement with AI

- AI-Powered Procurement: Gartner XPO is an AI-driven procurement platform that simplifies sourcing, supplier management, and contract negotiations. It offers a streamlined approach to procurement processes.
- Cost Optimization and Supplier Relations: The tool aids in optimizing costs and enhancing supplier relationships. Its AI capabilities enable smarter decision-making in supplier selection and contract management.
- Efficient Sourcing and Contracting: With Gartner XPO, businesses can efficiently manage their procurement lifecycle, from sourcing to contracting, ensuring the best value and terms in supplier agreements.

UiPath: Revolutionizing Operations with Robotic Process Automation (RPA)

- RPA Across Departments: UiPath stands out for its robotic process automation capabilities. It automates repetitive and time-consuming tasks across various departments, from finance to HR and IT.
- Boosting Operational Efficiency: By automating these tasks, UiPath not only improves operational efficiency but also frees up employee time. This allows staff to focus on more strategic, high-value work.
- Scalability and Flexibility: UiPath's platform is scalable and flexible, accommodating the changing needs of businesses. It integrates seamlessly with existing systems, enhancing the overall operational workflow.

BlackLine, NICE Actimize, Gartner XPO, and UiPath each play a pivotal role in transforming the landscape of finance and operations through AI. BlackLine automates essential accounting tasks, NICE Actimize provides advanced fraud detection, Gartner XPO streamlines procurement processes, and UiPath revolutionizes business operations with RPA. Together, these platforms demonstrate the power of AI to enhance accuracy, efficiency, and security in financial and operational processes, proving indispensable in today's fast-paced business environment

AI in Human Resources and Recruitment – Revolutionizing Talent Management

In the evolving field of human resources and recruitment, AI has become a game-changer. This chapter examines four innovative platforms - Workday Prism, Eightfold AI, BambooHR, and Hibob - each utilizing AI to transform various aspects of HR and recruitment processes.

Workday Prism: Insightful Analytics for Talent Management

- Employee Insight Analytics: Workday Prism harnesses AI and analytics to delve deep into data regarding employee engagement, retention, and performance. This allows HR professionals to make informed decisions in talent management.
- Guiding HR Strategies: The insights provided by Workday Prism are invaluable for shaping strategies around employee development, talent acquisition, and organizational growth.
- Enhanced Decision-Making: With data-driven insights, the platform aids HR leaders in making better decisions regarding talent utilization and employee development programs.

Eightfold AI: Streamlining the Recruitment Process

- AI-Driven Talent Acquisition: Eightfold AI brings efficiency to recruitment through its AI-powered talent acquisition platform. It automates candidate sourcing, screening, and interview scheduling, significantly speeding up the recruitment cycle.
- Improved Candidate Experience: The platform not only accelerates the recruitment process but also enhances the candidate experience, ensuring a smoother and more engaging hiring journey.
- Optimized Talent Matching: Eightfold AI's advanced algorithms enable precise matching of candidates to job roles, improving the quality of hires and reducing time-to-hire.

BambooHR: Comprehensive AI-Enhanced HR Software

- Automated HR Processes: BambooHR is an HR software platform that integrates AI-powered features to automate critical HR tasks like onboarding, payroll scheduling, and performance management.
- Streamlined Employee Interactions: The platform simplifies various HR

processes, making employee interactions more efficient and less time-consuming.

- Holistic HR Management: With BambooHR, businesses can manage the entire employee lifecycle from hiring to retirement, ensuring a cohesive and streamlined HR experience.

Hibob: Personalizing the Employee Experience

- AI-Powered Employee Personalization: Hibob leverages AI to tailor the employee experience within its HR platform. It offers automated employee recognition, feedback mechanisms, and opportunities for career development.
- Enhanced Employee Engagement: The platform's focus on personalizing the employee journey helps in boosting engagement, morale, and overall job satisfaction.
- Career Development Focus: Hibob's AI capabilities assist in identifying career development paths for employees, aiding in their professional growth and alignment with organizational goals.

Workday Prism, Eightfold AI, BambooHR, and Hibob each represent the innovative application of AI in the field of human resources and recruitment. From providing deep insights into employee dynamics to automating and personalizing the HR processes, these tools are redefining how organizations manage their most valuable asset – their people. Workday Prism offers analytic insights for strategic decision-making, Eightfold AI accelerates and refines the recruitment process, BambooHR streamlines essential HR tasks, and Hibob focuses on enhancing the employee experience. Together, these platforms demonstrate the transformative impact of AI in creating more efficient, engaging, and data-driven HR practices, essential for modern businesses navigating the complexities of talent management.

AI in Data and Analytics – Transforming Business Intelligence

In an era where data is king, the application of AI in data and analytics is revolutionizing how businesses extract, analyze, and interpret vast amounts of information. This chapter delves into four leading platforms in this domain: Tableau, Palantir Foundry, C3.ai, and Amazon SageMaker, each offering unique AI-driven capabilities to redefine business intelligence and analytics.

Tableau: Intuitive BI and Data Visualization

- AI-Powered Analytics: Tableau stands out as a Business Intelligence (BI) and data visualization tool, enriched with AI capabilities. It provides intuitive insights and recommendations, making

complex data easier to understand and use.

- Enhanced Decision-Making: The platform's ability to translate data into visually compelling and easily interpretable formats aids organizations in making better-informed decisions.
- Interactive Data Exploration: Tableau's interactive interface allows users to explore data in a dynamic and engaging way, fostering a deeper understanding of underlying trends and patterns.

Palantir Foundry: Advanced Data Analysis

- Deep Data Insights: Palantir Foundry offers a sophisticated data analysis platform that utilizes AI and machine learning. It excels in uncovering hidden patterns and predicting future trends from complex datasets.
- Informed Decision Making: The platform's robust analytical capabilities support data-driven decision-making, crucial for strategic planning and operational efficiency.
- Versatile Data Integration: Palantir Foundry's ability to integrate and analyze data from diverse sources makes it an invaluable tool for organizations dealing with large and varied data sets.

C3.ai: Predictive Analytics for Business Outcomes

- Forecasting Business Scenarios: C3.ai specializes in predictive analytics, using AI to forecast business outcomes. This foresight is pivotal for process optimization, risk management, and strategic planning.
- Cross-Industry Applications: The platform's versatile AI models are applicable across various industries, enabling businesses to predict and prepare for future scenarios and market dynamics.

- Operational Process Optimization: C3.ai's predictive capabilities facilitate the optimization of operational processes, enhancing efficiency and reducing risks.

Amazon SageMaker: Scalable AI and Machine Learning in the Cloud

- Comprehensive AI Platform: Amazon SageMaker is a cloud-based platform offering a full suite of tools and services for AI and machine learning. It supports the entire lifecycle of building, training, and deploying AI models.
- Scalability and Accessibility: SageMaker is designed for scalability, enabling businesses of all sizes to develop and deploy AI models efficiently on a large scale.

- User-Friendly Model Development: The platform provides an accessible environment for developing and fine-tuning AI models, even for those with limited machine learning expertise.

Tableau, Palantir Foundry, C3.ai, and Amazon SageMaker each play a crucial role in harnessing AI for data and analytics. Tableau's interactive data visualization, Palantir Foundry's in-depth data analysis, C3.ai's predictive insights, and Amazon SageMaker's scalable machine learning solutions collectively demonstrate the immense potential of AI in transforming the field of business intelligence. These platforms empower organizations to navigate the complexities of data, unlock valuable insights, and drive informed decision-making, positioning them at the forefront of data-driven business strategies in an increasingly data-centric world.

AI LITERACY AND THE FUTURE OF AI

Learning Outcomes:

By the end of this unit, learners should be able to:

- Understand the importance of AI literacy and its role in leveraging AI technology.
- Anticipate potential future developments and challenges in AI.
- Recognize the implications of AI on the job market and the need for workforce development in AI literacy.
- Apply this understanding to prepare for the impact of AI on their careers or businesses.

In the rapidly evolving landscape of Artificial Intelligence (AI), understanding AI literacy has become crucial. This unit delves into the significance of AI literacy, its impact on the job market, and the future developments in AI, particularly focusing on AI chatbots. By the end of this unit, learners will gain a comprehensive understanding of AI's role in modern technology and how to prepare for its impact on careers and businesses.

Understanding AI Literacy: Navigating the World of AI Chatbots

In the age of digital transformation, AI literacy has emerged as a crucial competency, extending far beyond the realm of traditional coding and programming skills. It represents a comprehensive understanding of how AI technologies, particularly AI chatbots, function, learn, and can be effectively implemented across various sectors. This writeup aims to provide insights into the multifaceted nature of AI literacy, emphasizing its importance in the current technological landscape.

Core Components of AI Literacy

1. Technical Understanding: At its foundation, AI literacy involves a basic grasp of the technical aspects of AI chatbots. This includes knowledge of algorithms, data processing, and machine learning techniques that enable chatbots to interpret and respond to human input. Understanding these elements is crucial for effectively deploying and managing AI chatbots in various applications.

2. Data Comprehension: AI chatbots are fueled by data. AI literacy, therefore, necessitates an understanding of data sourcing, management, and analysis. Recognizing how chatbots learn from data, including the process of training with datasets and the continuous learning from interactions, is vital. This knowledge helps in optimizing chatbot performance and ensuring they function as intended.

3. Ethical and Societal Implications: AI literacy also involves an awareness of the ethical considerations surrounding AI chatbots. This includes understanding the potential biases in AI algorithms, the ethical use of data, and the implications of AI decisions. It's about ensuring that AI chatbots are used responsibly, respecting privacy and avoiding harm.

4. Legal and Regulatory Awareness: As AI technology evolves, so do the legal and regulatory frameworks governing its use. AI literacy includes staying informed about these changes, especially regarding data privacy laws, AI patents, and compliance requirements. This knowledge is crucial for businesses to avoid legal pitfalls and maintain trust with their users.

5. Practical Application in Industries: AI literacy is not just theoretical; it's also about understanding the practical applications of AI chatbots in various industries. From customer service and healthcare to finance and education, AI chatbots have diverse applications. Understanding industry-specific requirements and challenges is key to leveraging AI chatbots effectively.

6. Limitations and Challenges: Finally, AI literacy involves recognizing the limitations of AI chatbots. This includes understanding the scenarios where human intervention is necessary and acknowledging the current technological constraints of AI.

AI literacy is a dynamic and multifaceted skill set that is becoming increasingly essential in a world where AI, especially AI chatbots, plays a significant role. By gaining a comprehensive understanding of AI chatbots, from their technical underpinnings to their ethical and practical implications, individuals and organizations can harness the full potential of this transformative technology.

Future Developments and Challenges in AI: A Focus on Chatbots

The trajectory of AI, particularly in the domain of chatbots, is on an unprecedented upward curve. With advancements in natural language processing (NLP) and machine learning (ML), chatbots are evolving to become more sophisticated and human-like. Let's explore the anticipated developments in AI chatbots and the challenges that accompany this rapid growth.

Anticipated Developments in AI Chatbots

1. Enhanced Natural Language Processing: Future AI chatbots are expected to exhibit significant improvements in understanding and processing human language. Advancements in NLP will enable chatbots to comprehend context, sarcasm, and complex queries more accurately, making interactions more natural and efficient.
2. Emotional Intelligence: The integration of emotional intelligence in AI chatbots is a promising development. Future chatbots might be able to detect subtle cues in human communication, such as tone and sentiment, allowing them to respond in a more empathetic and contextually appropriate manner.
3. Personalization: AI chatbots are likely to become more adept at personalization. By analyzing user data and past interactions, chatbots will offer more tailored responses, enhancing user experience and engagement.
4. Multimodal Interactions: Future chatbots may transcend text-based interactions, incorporating voice, facial recognition, and other sensory inputs to provide a more immersive and interactive experience.
5. Autonomous Learning and Adaptation: Advancements in ML will enable chatbots to learn autonomously from interactions and adapt to new scenarios without extensive reprogramming, thereby becoming more versatile and efficient.

Challenges Accompanying AI Developments

1. Ethical AI Practices: As AI chatbots become more advanced, ensuring ethical AI practices becomes paramount. This includes addressing concerns related to transparency, accountability, and the moral implications of AI decisions.

2. Privacy Concerns: With chatbots handling increasing amounts of sensitive data, safeguarding user privacy is a significant challenge. Ensuring compliance with data protection regulations and implementing robust security measures will be crucial.

3. Bias in AI Algorithms: AI chatbots are susceptible to biases present in their training data. Future developments must focus on creating unbiased, fair algorithms to prevent discriminatory practices and ensure equitable AI interactions.

4. Dependency and Overreliance: There's a risk of overreliance on AI chatbots, which might lead to a decrease in human skills and a dependency on automated systems. Balancing AI assistance with human capabilities will be a critical challenge.

5. Interoperability and Integration: As AI chatbots become more complex, integrating them with existing systems and ensuring interoperability across different platforms and technologies will pose a challenge.

The future of AI chatbots is marked by both exciting developments and formidable challenges. While advancements in NLP, emotional intelligence, and learning capabilities will enhance the effectiveness of chatbots, addressing ethical concerns, privacy issues, and algorithmic biases will be essential for responsible and sustainable AI growth. As we navigate this future, a balanced approach that leverages technological advancements while mitigating associated risks will be key to harnessing the full potential of AI chatbots.

AI's Implications on the Job Market: Navigating the New Landscape

The advent of Artificial Intelligence (AI), including AI chatbots, is reshaping the job market in profound ways. While there is apprehension about AI's potential to supplant human jobs, the reality is more nuanced. AI is altering the nature of work, creating new opportunities, and necessitating a shift in skills. Let's delve into the specific impacts of AI on the job market and the emerging trends in employment.

Transformation of Job Roles

1. Shift from Manual to Analytical Tasks: AI and automation are taking over repetitive, manual tasks, allowing human workers to focus on more analytical and creative work. This shift is leading to a demand for skills in problem-solving, critical thinking, and innovation.
2. Emergence of New Job Categories: AI is not just displacing jobs but also creating new ones. Roles like AI trainers, who teach AI systems how to mimic human behaviors, and AI ethicists, who ensure AI technologies are used responsibly, are becoming increasingly important.
3. Enhanced Job Efficiency: In many sectors, AI tools like chatbots are augmenting human work, leading to increased efficiency and productivity. For instance, in customer service, chatbots handle routine inquiries, allowing human agents to tackle more complex issues.

Demand for AI Skills

1. Technical Proficiency: There is a growing demand for professionals skilled in AI and machine learning, data science, and programming. These technical skills are essential for developing and managing AI systems.
2. Non-Technical Roles in AI: AI is also creating a need for skills in areas like supervision, maintenance, and decision-making. Understanding AI capabilities and limitations is crucial in these roles for effective management and strategic planning.
3. Interdisciplinary Expertise: The integration of AI in various fields is creating opportunities for professionals with interdisciplinary knowledge. For example, in healthcare, professionals who understand both medical science and AI applications are in high demand.

Reskilling and Upskilling

1. Continuous Learning: As AI technologies evolve, continuous learning and upskilling become essential. Workers need to stay abreast of the latest developments in AI to remain relevant in their fields.
2. Adaptability and Flexibility: The ability to adapt to new technologies and work environments is crucial. Flexibility and a willingness to embrace change are key traits for success in the AI-augmented job market.

3. Educational Initiatives: Educational institutions and businesses are increasingly offering courses and training programs in AI and related fields. These initiatives are crucial for preparing the current and future workforce for the AI-driven job market.

The implications of AI on the job market are multifaceted, involving both challenges and opportunities. While AI is transforming traditional job roles and creating new ones, it also necessitates a shift in the skill sets required in the workforce. Embracing continuous learning, adaptability, and interdisciplinary knowledge will be key for professionals to thrive in this evolving landscape. As AI continues to advance, staying informed and agile will be essential for navigating the new realities of the job market.

Preparing for AI's Impact on Careers and Businesses

The advent of Artificial Intelligence (AI) is not just a technological revolution; it's a paradigm shift in how careers are shaped and businesses are run. As AI continues to advance, its impact on various sectors is becoming increasingly profound. Preparing for this shift is essential for individuals and businesses alike. This 400-word writeup focuses on strategies to adapt to AI's growing influence in the professional world.

For Individuals: Enhancing AI Literacy and Skills

1. Continuous Learning: The first step for professionals is to embrace continuous learning. Keeping abreast of the latest developments in AI and related technologies is crucial. This can be achieved through online courses, workshops, and attending industry conferences.

2. Understanding Practical Applications: It's important to understand how AI is being applied in one's field. Whether it's AI in healthcare diagnostics, AI-driven analytics in finance, or AI-powered automation in manufacturing, knowing the specific applications helps in identifying the skills needed to work alongside these technologies.

3. Developing Relevant Skills: Beyond technical skills like programming and data analysis, soft skills such as critical thinking, problem-solving, and adaptability are increasingly important. Understanding AI's capabilities and limitations is crucial for effectively leveraging it in one's job.

4. Networking and Collaboration: Engaging with a community of AI professionals and enthusiasts can provide insights into emerging trends and job opportunities. Collaboration with peers can lead to a better understanding of how AI is transforming different sectors.

For Businesses: Integrating AI into Operations and Culture

1. **Strategic Planning with AI:** Businesses need to incorporate AI into their strategic planning. This involves assessing how AI can improve operations, enhance customer experiences, and create new business opportunities.

2. **Investing in Employee Training:** Companies should invest in training their workforce in AI literacy. This includes not only technical training for IT professionals but also general training for other employees to understand how AI impacts their work.

3. **Fostering an AI-Ready Culture:** Creating a culture that embraces innovation and technological change is key. Encouraging experimentation and learning about AI can help in smoothly integrating AI technologies into business processes.

4. **Ethical and Responsible Use of AI:** Businesses must also consider the ethical implications of AI. This includes ensuring data privacy, avoiding biases in AI algorithms, and being transparent about AI use with stakeholders.

5. **Collaboration with AI Experts and Vendors:** Partnering with AI experts and vendors can provide businesses with the necessary expertise and tools to implement AI solutions effectively.

Preparing for AI's impact is a multifaceted endeavor, requiring both individuals and businesses to stay informed, develop relevant skills, and embrace a culture of continuous learning and adaptation. By doing so, they can not only mitigate the challenges posed by AI but also leverage its potential to drive innovation, efficiency, and growth in their careers and business operations.

TO-DO

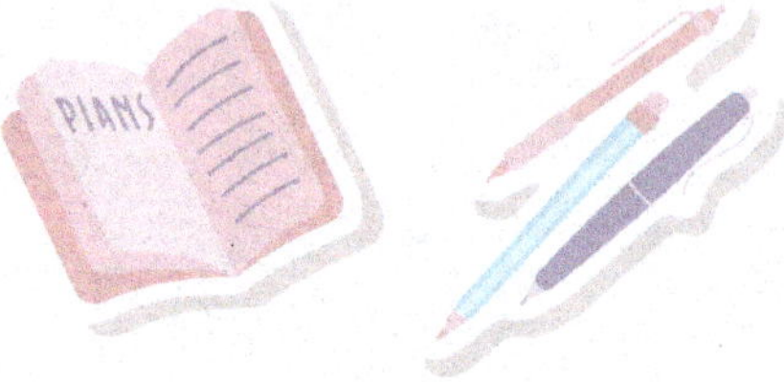

TO-DO

TO-DO

Made in the USA
Monee, IL
14 September 2025

25120643R00075

"Empowering Small Business with AI: A Practical Guide" is the essential roadmap for small and medium-sized business owners, including non-profits, looking to harness the power of artificial intelligence to elevate their operations. This comprehensive guide begins by unraveling the complexities of AI, presenting it in an approachable and applicable manner for those without a technical foundation.

The book dives into the innovative realm of AI chatbots, showcasing their versatility across a spectrum of industries. With engaging real-life case studies and vivid examples, it demonstrates how these digital assistants are revolutionizing customer engagement, operational efficiency, and personalized service in the world of commerce, healthcare, hospitality, and beyond.

Non-profit organizations will find particular value in the book's insights on leveraging AI chatbots to further their missions, engage with stakeholders, and optimize resource management. The practical advice offered within these pages serves as a blueprint for integrating AI solutions, overseeing their performance, and tracking their contributions to organizational goals.

"Empowering Small Business with AI" not only illustrates what AI chatbots can do but also provides a step-by-step strategy for their implementation. It is a vital tool for entrepreneurs and non-profit leaders seeking to stay ahead in an ever-changing digital marketplace.

ideal
Institute Of Technology

****

$29.99
ISBN 978-1-941862-50-6
52999>
9 781941 862506